Handbook for Water Distribution System Model Calibration

Handbook for Water Distribution System Model Calibration

American Water Works Association

Disclaimer

Senior Editorial Manager – Book Products: Melissa Valentine
Technical Editor: Suzanne Snyder
Copyright & Permissions Manager: Peggy Tyler
Publishing Operations Manager: Gillian Wink
Technical Illustrator: Michael Labruyere
Production: Innodata

Library of Congress Cataloging-in-Publication Data

Names: Parault, Patrick, author. | Ziemann, Sam, author.
Title: Handbook for water distribution system model calibration / by Patrick Parault and Sam Ziemann.
Description: Denver, CO : American Water Works Association,
   [2023] | Includes bibliographical references and index. |
Summary: "Hydraulic models are widely used to help solve engineering and operational problems in water distribution systems. To ensure that a model accurately represents a real system, model results are compared with physical measurements over a range of conditions. This book will help the modeler identify and mitigate the differences between field data and model results. It includes an introduction to calibration; suggested procedures; data requirements and collection guidelines; real-time modeling and continuous calibration; water quality calibration; and energy management issues"-- Provided by publisher.
Identifiers: LCCN 2023026273 (print) | LCCN 2023026274 (ebook) | ISBN 9781647171155 | ISBN 9781613006467 (ebook)
Subjects: LCSH: Water--Distribution--Computer simulation. | Hydraulics--Data processing. | Expert systems (Computer science)--Validation.
Classification: LCC TD481 .P37 2023 (print) | LCC TD481 (ebook) | DDC 628.1/440113--dc23/eng/20231115 LC record available at https://lccn.loc.gov/2023026273 LC ebook record available at https://lccn.loc.gov/2023026274

Printed in the United States of America

ISBN: 978-1-64717-115-5
ISBN, electronic: 978-1-61300-646-7

American Water Works Association
6666 W. Quincy Avenue
Denver, CO 80235-3098
303.794.7711

CONTENTS

ACKNOWLEDGMENTS

The AWWA Technical & Educational Council, the Engineering Modeling Applications Committee (EMAC), and Model Calibration Subcommittee gratefully acknowledge the contributions made by those volunteers who drafted, edited, and provided the significant and critical commentary essential to developing the Model Calibration Handbook. Many hours were spent in the preparation of this first edition of the Handbook to ensure the overall technical quality, consistency, and accuracy.

EMAC Committee Chairs
Melissa Brunger, Freese and Nichols, Dallas
Christie Patel, New Alexandria, Va.

Reviewers
Patrick Parault, AKRF Inc, N.Y.
Sam Ziemann, C3 Water, Kitchener, ON
Michelle Dutt, Garver, Austin, Tex.

Authors
Tom Walski, Bentley Systems, Nanticoke, Pa.
Sasa Tomic, Burns and McDonnell, N.Y.
Eric McLeskey, Carollo Engineers, Phoenix, Ariz.
Aurelie Nabonnand, Carollo Engineers, Seattle, Wash.
Matt Huang, Carollo Engineers, Portland, Ore.
Jim Cooper, Arcadis, Akron, Ohio
Rajan Ray, Trinnex, Boston, Mass.
Luke Butler, Qatium, Oakville, ON
Lindle Willnow, AECOM, Chelmsford, Mass.
Walter Grayman, WM Grayman Consulting, Oakland, Calif.
Ferdous Mahmood, City of Dallas
Meg Roberts, Hazen and Sawyer, Greensboro, N.C.

CHAPTER 1

Introduction to Model Calibration

Tom Walski, Sasa Tomic

Water distribution system hydraulic models are widely used in the planning, design, and operation of water distribution systems. Hydraulic models consist of two major components: the numerical solver, which solves the hydraulic and water quality equations, and the data that describe the particular system as input to the hydraulic solver.

Since the advent of digital computers, numerical solvers have become progressively more powerful and reliable. However, the accuracy of the model depends on the extent to which it reflects the real system. To ensure that the model is an accurate representation of the real system (essentially a "digital twin" of the real system), it is necessary to compare model results with real-world data and adjust model parameters to reduce the level of uncertainties in the model results, and then identify and correct as many discrepancies as is feasible.

According to the Calibration Subcommittee of the AWWA (American Water Works Association) Engineering Modeling Applications Committee (2013a), "water distribution model calibration consists of comparing model results with accurate field measurements and making adjustments to model and reviewing field data to improve agreement over a range of conditions that correspond to how the model is intended to be used." Calibration should result in a more accurate model, a better understanding of the strengths and weaknesses of the model, and greater confidence in model results.

Hydraulic models of water distribution systems are widely used to help solve engineering and operational problems in water systems. To ensure that a model accurately represents a real system, model results are compared with physical measurements over a range of conditions. If the model predictions agree with the data, the model can be used with confidence for the conditions it was calibrated.

No model perfectly matches field conditions. This is due to the issues described in the following table. There are many sources of differences that are not always obvious. This book will help the modeler identify and mitigate the differences between field data and model results.

Issues	Examples
■ Simplifications	■ Model skeletonization
■ Approximations	■ Distribution of unmetered water to demand nodes
■ Assumptions	■ Summer demand pattern is similar to that of spring
■ Inaccuracies	■ Use of nominal diameters instead of actual internal diameters
■ Model errors	■ Wrong connectivity at complex intersections
■ System description errors	■ Closed valves that should be open
■ Data collection errors	■ Inaccurate sensors or incorrect data handling
■ Natural system variability	■ Accounting for day-to-day system changes

WHY CALIBRATE?

Water distribution models have become essential tools for system planning, design, and operation. Uncalibrated models almost always contain errors because of the issues described above.

Depending on the system and the intended model use, calibrating a model can consume significant resources and it is often tempting to shortcut the process. However, decisions based on the model results, such as construction projects and operational decisions, can result in large costs. The cost of preparing a well-calibrated model is minuscule compared with the potential for improved decision-making. In general, better calibration leads to better models and better understanding of the model accuracy, which in turn leads to better decisions. Without validation, it is difficult to determine model accuracy and to decide if decisions have been based on an appropriate model.

Water distribution models can be used for a myriad of purposes including the following:

- system planning
- main extensions
- new land development
- fire flow studies
- criticality studies
- pressure zone and district metered area analysis
- pump selection

- tank locations and sizing
- pump and valve controls
- emergency planning
- emergency response
- operator training and shut-down planning

The need and extent of model calibration will differ depending on the purpose of the model. As such, there can be no single process for calibrating a model. Calibrating a 100,000-pipe water quality model is much different than calibrating a fire flow model in a 200-pipe rural system.

Table 1-1 Sources of error

Physical
Pipe size/location
Pipe connectivity
Pipe roughness
Pressure zone boundary
Pump curves
Pipe material/age in GIS
System changes since model built
Elevation data
Operational
Valve open/closed/throttled status
Control valve operation/settings
Transient events
Actual operations not matching control rules
Unusual operations when data were collected
Tank water levels
Pump status/speed
Lack of sufficient sensors/gages
Water quality reaction rates
Demands
Spatial allocation
Model does not reflect conditions when data is collected
Large customers with atypical demand patterns
Not accounting for seasonal changes in demand
Data
Inaccurate/uncalibrated gages/meters
"Latched" data from SCADA (Supervisory Control and Data Acquisition)
Understanding SCADA data – average vs. instantaneous

Calibration is not easy. As will be seen in Chapter 4, the number of reasons why calibration may not be adequate are many and can include:

Calibration should be approached in a logical procedure. While some random sensitivity analysis can provide insights, modelers need to approach the work with confidence that the fundamental hydraulic principles apply and that the reason for the discrepancy between model results and the observed data generally lies in the model input or observed data.

Most hydraulic models require some calibration for even basic uses, and numerous model adjustments are often required (Walski, 1996). However, model developers and users must understand that adjusting the wrong parameters to achieve better model agreement results in compensating errors (Walski, 1983). Guidance in determining the source of model errors is scarce. For example, using an unreasonably high C-factor can compensate for incorrectly overestimated demands. The number of possible adjustments often overwhelms the modeler.

One of the worst things a modeler can do is assume that a discrepancy between the model results and observed data is due to a single type of error and only adjust corresponding parameters (e.g., pipe roughness). Adjusting the wrong parameter can lead to "calibration by compensating errors" that can make the model look accurate for a set of calibration data but can introduce errors that will plague the model for other applications.

Model calibration can also reveal surprises. Modelers can find valves that were mistakenly closed, pumps no longer operating on their curves, errors in GIS data, locations with excessive disinfectant decay, or inaccurate sensors. Exposing these issues can have immediate benefits for the water utility beyond having a better model.

The calibration effort should also lead to improved communication between modelers and operations personnel. Once non-modelers see the value of models and understand their limitations, they will be better able to leverage the model results in their work with confidence.

FIELD DATA REQUIREMENTS

While all field data may be useful for calibration of any type of water distribution model, the relative importance of each type of data depends on the application of the model. The table below shows which type of data is most important for each application.

For example, fire hydrant flow tests are important for calibrating capacity-based planning models and for service extension planning, or in other situations where performance under high demands is critical, while such data are less important in an energy study where performance on extended periods on typical days is usually more important. Meanwhile, energy studies rely heavily on pump curve tests. This is summarized in Table 1-2, where a high number indicates that a type of calibration data is more important for that type of model application than other types of data.

Differences between a model and field observations could stem from poor field data that needs to be corrected. (Here the term *field data* includes manual measurements, SCADA system values, downloads from data loggers, and records from customer information systems.) But if the data are considered valid, the model inputs must be modified to minimize or at least understand any differences.

Even with valid field data, modelers need to understand level of accuracy. Some data are very helpful for model calibration (e.g., a hydrant flow test far from the source), but others provide minimal insights (e.g., pressure reading near an elevated tank where the water level is known) (Walski, 2000).

Table 1-2 Types of model application

Type of Calibration Data	Type of Model Application							
	Rural System Planning	Medium System Planning	Large System Planning	Main Extensions	Transmission Mains	Operational Studies	Energy Studies	Water Quality Studies
System Pressure and Flow Measurement	3	2	2	2	2	1	1	1
Fire flow/Roughness Test	1	3	2	3	1	2	1	2
Time Series Data (SCADA/loggers)	2	2	3	1	3	3	2	2
Pump Curve Test	2	2	2	1	2	3	3	1
Tracer Test	1	1	1	1	2	2	1	3
Constituent Measurement	1	1	1	1	1	2	1	3
Energy Measurement	1	1	1	1	2	3	3	1

TYPES OF CALIBRATION

There are multiple situations where model calibration is often performed:

1. At the completion of initial model building
2. Immediately before use of the model for a particular problem
3. In-real time modeling where the model is run frequently.

In the first situation, the potential model uses are open-ended and hence calibration requirements are open-ended. The utility needs to define the model use expectations so that model calibration targets can be defined. Especially when the model is being developed outside of the water utility, it is important to clearly define the use of the model and the metrics used to assess calibration, realizing that the more stringent the targets, the greater the resources needed to achieve them. While it is good to have numerical calibration targets, they may need to be adjusted as the water system, utility's priorities, available data, budget, and the hydraulic model change.

In the second situation, the use of the model is well understood, and the extent of calibration is clearer. If the model is to be used for modeling a new subdivision, the model accuracy can be evaluated in terms of hydraulic capacity in that part of town, for example, by comparing model results with fire flow tests at the connection point to the existing system.

In the third situation, the model can be almost continuously compared with data provided by the SCADA system. When the model and field data diverge, there is an opportunity to understand the cause and improve the model (or identify inaccurate sensor input). However, these comparisons can only be made at sensors connected to the SCADA system which typically are spatially sparse.

DEVELOPMENT OF CALIBRATION PROCEDURES

In the earliest days of digital models, users were focused primarily on building the models and replacing the process of manual calculations. As the numerical solvers became more powerful, easy to use, and reliable, the emphasis shifted to the extent to which the models matched field conditions.

With early models, merely matching static pressures in the model was considered an indication of success (Eggener and Polkowski, 1983). While such comparisons could detect gross errors in connectivity or boundary conditions, there could still be significant errors when systems became stressed. Later work emphasized the need to achieve significant head loss to ensure that models were calibrated over a range of conditions and hydrant flow tests became more widely accepted as a source for the model calibration data (Walski, 1984). Techniques for water quality modeling evolved later (Grayman, Clark, and Males, 1988; Walski 2017). Every edition of AWWA Manual M32 on Distribution Modeling has had some discussion of calibration, starting with a few pages in the 1989 edition and evolving into an extensive chapter in later editions (AWWA, 1989).

There is considerable guidance in the literature on model calibration including AWWA (2012), Edwards, Cole, and Brandt (2006), Hirrel (2008), Ormsbee and Lingireddy (1997), Speight et al. (2010), Tomic (2015), Walski (1983), Walski (1990), and Walski et al. (2003). A literature review by the AWWA Calibration Subcommittee (2013) found more than 200 papers published about calibration. While these papers provide useful information, many of the solution methods included the hidden assumption that the modeler knew the source of error in the model.

Most research on calibration has focused on determining how to adjust model parameters based on the assumption that all errors were concentrated in a few parameters, such as pipe roughness and demands. Numerous researchers have applied a variety of optimization tools to automate calibration which has been successful to the extent the procedure is adjusting the correct parameter(s) (for example, Lansey and Basnet, 1991; Savic and Walters, 1995).

Acceptance of Model

One of the most common questions that is asked about model calibration after "Where do I start?" is "When are we done?" The answer, unfortunately, is "never." Water systems are dynamic; new facilities are added, and operating procedures change constantly. Nevertheless, modelers want to know when they can begin using a newly built or recently upgraded model, and this involves setting some system-specific and use-dependent targets for calibration. All models are approximations to the real world but can be used even when they are not perfect. Understanding the differences between the model and field data gives the user an appreciation for the extent to which the results can be used to support decision-making.

While it is desirable to have numerical targets, for a given model, situations will arise where these prove to be too stringent or lenient once calibration is underway. It is difficult to develop targets before the initial attempts at calibration have been completed. *The end point of calibration should be when the benefits of additional calibration no longer exceed the cost.*

It is highly unlikely that any model will be perfectly calibrated when it is first created. When the calibration of a model is not adequate for its intended use, adjustments must be made. It is essential that before adjusting the model, the modeler understands why a certain parameter is being adjusted. Adjusting the wrong parameter (e.g., changing pipe roughness when an incorrect elevation caused the model calibration to be poor) may result in a model that may initially look calibrated but in other situations will not give acceptable results because it has been calibrated by compensating errors. The goal of calibration is to have the model faithfully reproduce the system, not simply match a handful of data points.

Essentially, there are two steps to the calibration process.

1. Determining why there are differences between the model and field data.
2. Making the necessary adjustments to achieve calibration

As will be seen in later chapters, the first step is the more difficult as it involves intuition and detective work as well as hydraulic expertise.

There is no way to give a general "Yes/no" answer to the question, "Is this model calibrated?" Calibration can only be evaluated on a continuous scale from poor to excellent for a given intended use. Calibration can only be judged on the model's suitability for specific tasks.

A model is a decision support tool and not an end in itself. There are two roles in modeling:

1. Modeler who builds the model and performs calibration comparisons
2. Decision-maker (engineer/operator/planner) who relies on model results to help make decisions.

These roles can be filled by any combination of two individuals (or teams) in the water utility, two separate consultants to the utility or one individual (or team) that performs both tasks. Extensive, frequent communication between modelers, system operators and consumers of model results is necessary for successful calibration.

In deciding if calibration is adequate, the decision-maker or end user should not ask the modeler if the model is calibrated but instead, the modeler should show the decision-maker what was done for calibration, and the decision-maker and modeler can decide if the model is sufficiently calibrated for a particular task. The decision-maker and modeler can then assess the need for additional calibration work. A model may be well calibrated for one task but not for another. (For example, a model may be well calibrated for fire flow analysis at a location but poorly calibrated for a system-wide water quality analysis). At some point, a decision must be made that the model is adequately calibrated for a specific task for which it will be used.

While it has not been proven feasible to provide general numerical guidelines for the acceptability of model calibration, a modeler and a decision-maker can reach an agreement on targets for model calibration for a particular task (Walski, 2019). The decision on the acceptability of model calibration depends on

1. Uses of the model
2. Sensitivity of the decisions being made to model accuracy
3. Quality and availability of field data
4. Metrics used to evaluate calibration
5. Budget, resource, and time constraints.

Even in a generally well-calibrated model, there may be areas where the model and the field data do not agree, and the utility may not have the resources to solve the problem. However, the model still may be a very useful decision support tool if the users understand its limitations. For example, a model may have been well calibrated in an area where growth in the system is expected and hence be very useful for evaluating capacity for land development. The same model may have inaccurate pump efficiency vs. flow data and would do a poor job of assessing energy use in that system.

Calibration may also be limited by the accuracy of field data. It is important to know the accuracy of such data. For example, it is not reasonable to expect hydraulic grade elevation to match within 2 ft (1 psi) if the elevation of the pressure gage is only accurate to ± 10 ft. Similarly, model results should only be reported to the precision of the data used in calibration. (For example, if pressure is measured to ± 1 psi, the model results should not be reported with more significant digits, such as 65.253 psi.)

Boundary conditions (e.g., pump status, PRV settings) in the model should be known to the extent possible. Good calibration should test a model at times and locations where the model is more likely to be inaccurate. For example, if the chlorine concentration at the source is known exactly, the model and field measurements should not only be compared close to the source but also at more remote locations. Similarly, pressures should be compared at times when there is significant head loss in the system (i.e., when the system is being stressed due to high flows) unless the purpose is to only check elevations and boundary heads.

While a model with limited calibration may provide some insights into system performance (and a model of a proposed system can't be calibrated), confidence in the model increases with the extent of calibration. A properly calibrated model is an important asset for water utility. A utility needs to commit sufficient resources to the task of calibration.

A calibrated model may need to be updated as the system changes due to new infrastructure, modified operation, or changes in water consumption. *Calibration is never fully completed.*

ANNOTATED BIBLIOGRAPHY

The AWWA Calibration Subcommittee of the Engineering Modeling Application Committee (EMAC) developed an annotated bibliography of model calibration papers and presentations with several hundred entries (AWWA Model Calibration Subcommittee, 2013a). The annotated bibliography can be found at the EMAC Resource web page: https://www.awwa.org/Resources-Tools/Resource-Topics/Engineering-Modeling-Applications.

Most of the papers can be categorized as either case studies for specific systems or research studies on parameter tuning using optimization. Only a handful of entries provide general guidance on calibration.

OVERVIEW OF THIS HANDBOOK

The following chapters provide detailed instructions on water distribution system model calibration as described below.

Chapter 2 provides an overview of the calibration process with an emphasis on how to approach calibration in a logical manner.

Chapter 3 describes the type of data needed to perform calibration.

Chapter 4 gives detailed steps for determining which parameters may need to be modified to achieve calibration and making the best calibration adjustments.

Chapter 5 describes additional tools and issues for model calibration when real-time data are available.

Chapters 6 and 7 extend the concepts from the previous chapter for models used for water quality and energy analysis.

This book has been written with the expectation that the reader is familiar the basics of water distribution hydraulics and familiar with modeling terminology as described in publications such as AWWA Manual M32 (AWWA, 2012).

REFERENCES

AWWA, 1989, "Distribution Network Analysis for Water Utilities," AWWA Manual M32 (first of multiple editions), Denver, Colo.

AWWA Model Calibration Subcommittee, 2013a, "Committee Report: Defining Model Calibration," *Journal AWWA*, 105(7), p. 60–63, July.

AWWA Model Calibration Subcommittee, 2013b, "Hydraulic Model Calibration: What Don't We Know?" AWWA Distribution Symposium, Denver, Colo.

AWWA, 2012, *Computer Modeling of Water Distribution Systems*, AWWA Manual M-32, Denver, Colo.

Edwards, J., Cole, S., and Brandt, M., 2006, "Quantitative Results of EPS Model Calibration with a Comparison to Industry Guidelines," *Journal AWWA*, 98(11), p. 72–83, November.

Eggener, C.L. and Polkowski, L. 1983, "The Network Model and the Impact of Modeling Assumptions," *Journal AWWA*, 106-TE5, p. 189, April.

Grayman, W., Clark, R. and Males, R., 1988, "Modeling Distribution System Water Quality: Dynamic Approach," *Journal of Water Resources Planning and Management*, 114-3, p. 295, May.

Hirrel, T.D., 2008, "How Not to Calibrate a Hydraulic Network Model," *Journal AWWA*, 100(8), p. 70–81, August.

Lansey, K. and Basnet, C., 1991, "Parameter Estimation for Water Distribution Networks," *Journal of Water Resources Planning and Management*, 117-1, p. 126, January.

Ormsbee, L. and Lingireddy, S., 1997, "Calibrating Hydraulic Network Models," *Journal AWWA*, 89(2), p. 42–50, February.

Savic, D. and Walters, G., 1995, "Genetic Algorithm Techniques for Calibrating Network Models," Report No. 95/12, Center for Systems and Control Engineering, University of Exeter, Exeter, UK.

Speight, V. et al., 2010, *Guidelines for Developing, Calibrating and Using Hydraulic Models*, Water Research Foundation, Denver, Colo.

Tomic, S, 2015, "Steady State Model Calibration," in Model Calibration Workshop at *AWWA Water Infrastructure Conference*, Denver, Colo.

Walski, T.M., 1983, "Technique for Calibrating Network Models," *Journal of Water Resources Planning and Management*, 109(4), p. 360–371, October.

Walski, T.M., 1984, *Analysis of Water Distribution Systems*, Van Nostrand Reinhold, New York, N.Y.

Walski, T.M., 1986, "Case Study: Pipe Network Model Calibration Issues," *Journal of Water Resources Planning and Management*, 112(2), p. 238–249, April.

Walski, T.M., 1990, "Sherlock Homes Meet Hardy Cross or Model Calibration in Austin, Texas," *Journal AWWA*, 83(3), p. 34–38, March.

Walski, T.M., 2000, "Model Calibration Data: the Good, the Bad and the Useless," *Journal AWWA*, 92(1), p. 94–99, January.

Walski, T.M. 2015, "Calibration Basics," in Model Calibration Workshop at *AWWA Water Infrastructure Conference*, Denver, Colo.

Walski, T.M. 2017, "Procedure for Hydraulic Model Calibration," *Journal AWWA*, 109(6), p. 55–61, June.

Walski, T.M. 2019, "Why Global Standards for Calibration of Water Distribution Models Won't Work," *Journal AWWA*, 111(5), p. 16–19, May.

Walski, T.M. et al., 2003, *Advanced Water Distribution Modeling and Management*," Bentley Systems, Exton, Pa.

CHAPTER 2

Procedure for Hydraulic Model Calibration

Tom Walski, Sasa Tomic

Virtually every water distribution system (WDS) model requires some level of calibration adjustment so that the model accurately represents the behavior of the real water system. The modeler is faced with an overwhelming number of potential adjustments to achieve calibration (see Table 1-1). The procedure presented in this book provides a logical way to determine what parameters need adjustment so that modelers can calibrate their models efficiently, without being misled by compensating errors.

Modelers have used a wide variety of approaches to calibrate models. There is no single "right way" to calibrate a model. However, there is a wide range of pitfalls awaiting the modeler. This chapter presents a logical process for modelers to proceed through a calibration study efficiently. It is not the only approach. A typical model uses and lack of desired data can necessitate variations from the procedure described below. Nevertheless, attempting to follow the procedure should lead to acceptable results with a minimum of problems.

A significant problem facing modelers during calibration is figuring out where to start when the number of changes can be daunting. Since numerous adjustments can be made, the modeler is often faced with too many choices of what to adjust. What a modeler would like to do is pick a manageable subset of parameters to adjust, verify them, and then move on to the next set of calibration data. This book provides an orderly approach for modelers to follow during the calibration process to avoid the kinds of errors already discussed, not trying to solve for all inputs at once, which is generally impossible.

Table 2-1 Types of discrepancies that can be found with different types of model runs

Type of Model Comparison	Differences That can be Identified
Steady, low flow	Pressure zone boundaries, elevations, inaccurate sensors, tank levels, pump curves, PRV settings, pump status
Steady high flow (hydrant test)	Pipe roughness, closed valves, demands, pipe connectivity
EPS with known pump/valve status	Demand patterns, pump curves, valve head loss
EPS testing controls pump valve status	Pump and valve controls
Water quality	Source quality, reaction rates, tank mixing
Energy	Pump efficiency, energy tariff

EPS—extended period simulation, PRV—pressure reducing valve

THE PROCEDURE

A modeler should try to limit the number of parameters that are in question at each step of calibration so that the source of any errors can be identified and corrected. While following the steps in Figure 2-1 is not required, skipping one of the steps may save a little time upfront but is likely to make later steps much more difficult.

The terms *macro* and *micro calibration* (Ormsbee and Lingireddy, 1997) have been used to describe this type of approach, but those terms are subject to interpretation. By precisely describing the type of data and nature of adjustments at each step, as shown in Table 2-1, the procedure in this chapter should reduce uncertainty. The procedure presented in this chapter is based on Walski (2017).

The photos in Figure 2-1 through 2-4 illustrate some of the kinds of facilities that can be accessed in calibration as it is not possible to readily see inside pipes. In Figure 2-1, using data when the hydraulic grade line is relatively flat provides a way to identify errors in pressure zone boundary valves. When the slope of the hydraulic grade line is increased as in hydrant flow tests (Figure 2-2), it is possible to identify closed or partly closed valves.

The overall goal of calibration is to adjust the model outputs to better match valid field data. However, at any stage in the procedure, inconsistent data should be challenged and discarded if there are doubts about its quality. See Chapter 3 for a more detailed discussion on this topic.

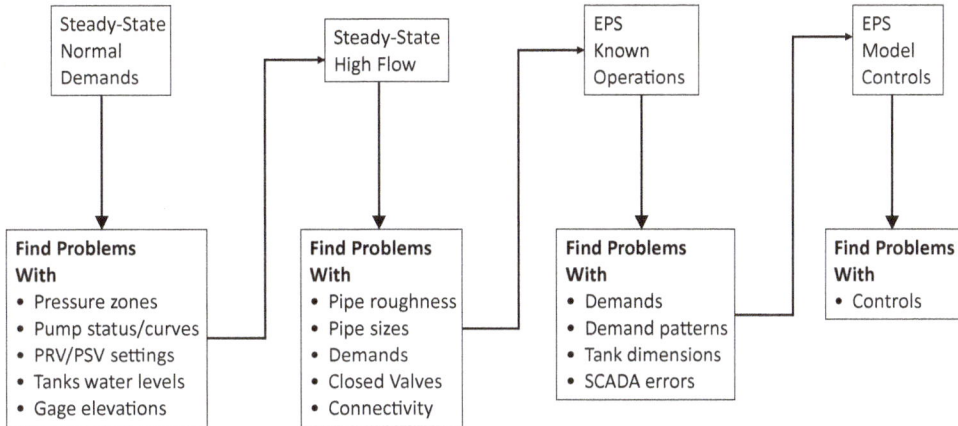

| Steady-State Normal Demands | Steady-State High Flow | EPS Known Operations | EPS Model Controls |

Find Problems With	**Find Problems With**	**Find Problems With**	**Find Problems With**
• Pressure zones	• Pipe roughness	• Demands	• Controls
• Pump status/curves	• Pipe sizes	• Demand patterns	
• PRV/PSV settings	• Demands	• Tank dimensions	
• Tanks water levels	• Closed Valves	• SCADA errors	
• Gage elevations	• Connectivity		

Source: Courtesy of Tom Walski

Figure 2-1 Suggested workflow for model calibration

Source: Courtesy of Tom Walski

Figure 2-2a Check status of isolation valves

Matching tank level fluctuations in the model and the field data provides insights into demand patterns and system operation. Tanks (Figure 2-2c) are the only places where it is possible to actually see the level of the hydraulic grade line.

Problems with pump curves can be detected even during low flow conditions while problems with pump operations can only be found during extended period model runs (Figure 2-2d).

Source: Courtesy of Tom Walski

Figure 2-2b To make it easier to identify problems with roughness and incorrect connectivity, hydrant flow tests can provide increase velocity and head loss.

Source: Courtesy of Tom Walski

Figure 2-2c Matching tank water level fluctuations is a desired way to find errors in demand patterns.

Source: Courtesy of Tom Walski

Figure 2-2d Pump controls can be checked using EPS runs where pumps are not forced to follow a time pattern.

Steady-State Normal-Flow Conditions

During normal and low-flow periods in most water distribution systems, the hydraulic grade line (HGL) is relatively flat because the system's low velocities result in small head losses. This means that the effects of pipe roughness, demands, and closed valves on HGL are small, if measurable at all. During normal low-flow conditions, the value of the HGL is primarily determined by water levels in tanks or by pump and/or PRV discharge and the system demands. In general, it is not recommended to start model calibration by adjusting pipe roughness.

In this case, the HGL should not vary much across the zone and a model should match field data. If it does not, the most likely sources of error are as a subset of Table 1 in Chapter 1:

- Incorrect tank water level
- Incorrect PRV settings or reference elevation
- Incorrect pressure zone boundaries
- Report the calculated HGL level based on model node level (usually ground level) when pressure sensors at located at a different elevation (e.g., nearby hydrant outlet level)
- Incorrect pump status/speed
- Incorrect node elevations
- Differing demands at time measurements from demands loaded into model

Comparisons between a model and field data should be made in terms of HGL, rather than pressure. While they are related, HGL values can be compared at a glance, whereas, unless the ground is completely leveled, pressures cannot be compared so easily. Figure 3 shows how HGL data can provide more insights than pressure data. For example, where the tank HGL may be 650 ft, HGL's values in the system that are close to the tank HGL (such as 645 or 649) can easily be identified while different HGL values, such as 693 ft, would stand out immediately as an error or an anomaly in the system. Similarly, HGL values above the tank value could indicate a source of water that is filling the tank or a different error. HGL values below the tank water level would indicate flow from the tank. With pressure data, such a problem cannot be seen without first correcting for changes in elevation which could be incorrect themselves.

Locations where incorrect pressure zone boundaries are suspected can be easy to detect with HGL measurements near the boundary. Attempting to adjust pipe roughness to compensate for incorrect boundary head leads to an inaccurate model.

A comparison of pump discharge HGL is important but can only show if a single point on the pump curve is correct and not the entire curve. If there are questions about the curve, the curve can be verified in a separate test (see Chapter 3).

Steady-State High-Flow Conditions (System Under Stress)

During normal flows when head loss is low, HGL comparisons cannot precisely reveal problems that occur due to head loss between the boundary nodes and pressure measurement locations. During high-flow conditions at peak demands, discrepancies that could not be detected when the head loss was low become obvious. The most likely sources of error are

- Closed valves
- Pipe roughness
- Demand locations
- Demand magnitude
- Network connectivity
- Pipe diameter

The key with steady-state calibration tests the high flow to ensure that the head loss between the source and the measurement location is significantly greater than the error in measurement. For example, if the HGL can be reliably measured to within +/- 5 ft, then the head loss should be on the order of 25 ft or greater.

Sufficiently high head loss can occur in some systems during normal demands, but more often, these high flows need to be induced by the operators. For smaller mains, the velocity and head loss are usually too low such that it is best to increase the velocity in the pipes by conducting fire hydrant flow tests (as described in Chapter 3). When hydrant flow tests are conducted, it is important to know the boundary heads/pressures (i.e., tank levels, PRV settings, pump status/discharge pressure) at the time of the test. In large transmission mains, sufficient head loss can be achieved by measuring head loss over long distances during peak demand times. In following the recommended approach, modelers should try to reduce possible sources of error so those that are candidates for modification are manageable.

The usual sources of error under high-flow conditions are closed valves, incorrect connectivity, wrong overall demands, or incorrect roughness. If the errors in HGL are somewhat uniformly distributed across a pressure zone, the errors are usually due to roughness or demands. Demands can be verified by determining the net flow into the pressure zone at the time HGL readings are made. The range of variation of roughness can be constrained by conducting C-factor tests on a range of different types and ages of old pipes (see Chapter 3). (C-factors are generally known for new pipes). If there are a great deal of old, unlined cast iron pipes, the initial roughness adjustments are usually based on pipe age. In systems with mostly plastic or lined pipes, HGL is not very sensitive to roughness adjustments, unless the roughness is extensive (e.g., 100-year-old unlined cast iron pipe) or there ae some unusual sources of roughness.

If discrepancies between measured and modeled HGL values only occur in a few locations and the model predicts HGL higher than measured, the primary source of error is usually closed valves. These can be identified by using multiple residual pressure gages or data loggers during hydrant flow tests. Locations upstream of the closed valve will show a very low pressure drop during the test while those downstream will show a significant, abrupt drop in the region of the closed valve.

It is best to collect HGL data far from tanks and other known boundary locations (e.g., PRVs). Conducting a test near a tank only provides information about what is occurring between the tank and the measuring point and only minimal insight about what is occurring beyond that point.

Incorrect pipe diameters can also best be identified by induced high flows, although these errors are not common. However, the data are usually not sufficiently precise to identify differences between nominal and actual pipe diameters. Typical high flows can be effective in identifying modeled diameter errors in transmission mains.

Once the modeler has corrected the model and the field data through these first two steps, the modeler should feel confident that the model can be relied upon for most design problems where steady state analysis is sufficient.

EPS—Known Control Status

Once a steady-state model is adequately calibrated, the next step is an EPS calibration. A steady-state model may correctly identify a pump or valve's initial status or settings, but an EPS model must simulate these properties over time. Calibration parameters that can be adjusted in EPS calibration include demand patterns and changes in the settings for pumps and valves. Comparisons at this step usually center around matching flow rates and tank water level fluctuations in the model with data from the Supervisory Control and Data Acquisition (SCADA) system. If earlier steps were done correctly, HGL or pressure should not be the most important parameters to judge the model, although if earlier steps were unable to adequately identify sources of poor HGL or pressure calibration, an EPS may help to identity those sources.

There are two kinds of differences that surface during this step:

1. Tank levels in the model moving in the wrong direction from the SCADA data
2. Tank levels moving in the correct direction but at the wrong rate.

The first is usually due to problems with pump controls while the second is due to inaccurate diurnal demand patterns. These two problems should be separated and solved one at a time.

Initially, the uncertainty in pump controls can be removed from the problem by using time-based controls for the known historical period. With this source of uncertainty removed, the modeler can focus on the demand patterns. At this stage, flow metering data is much more useful than pressure data. If flow meters are available for pressure zones or at the district meter area level, they can help guide adjustments.

Poor model agreement with field data can be caused by differences between the demands loaded in the model (usually average day) and those from the day on which calibration data were collected. There may have been a special event that day, e.g., the local university may have been on spring break, or there is a large sporting/music event. Using the daily production to scale demands up or down can usually account for demand differences, although an event such as a large fire may need to be placed at a specific location. Preferably, a typical day that resembles the average modeled conditions should be used for initial EPS calibration. However, confidence in the model grows with the number of days for which comparisons are made successfully.

Models with time-based controls should only be used for the specific day (or time period) for which it was calibrated. Extrapolating the model beyond that period is not reliable because each day is different to some extent. Even if there isn't a special event, there are always fluctuations in water use from one day to the next, and demands for the "typical day" that were loaded into the model are very rarely repeated.

The increased adoption of "live modeling/digital twins" makes it possible to quickly and frequently compare model results to the current day or any day in the SCADA historian. EPS calibration is also being changed by the ability to use Automated Meter Reading and Advanced Metering Infrastructure to understand individual customer water use at a much more granular level as a basis for developing demand patterns. However, precise leakage and other non-metered usage values are usually elusive.

EPS—Unknown Control Status

While using time-based controls to simulate a specific day may be adequate to check demand patterns, the model must be used for days that differ from the day for which calibration data were collected. To deal with these variations, hydraulic models usually contain rule-based control statements.

Comparing variations in water level, flow or pressures over time can often lead to poor initial calibration because rule-based controls may frequently be overridden by operators. Nevertheless, graphs of these properties showing field and model results should be reasonably close. Otherwise, the rule-based controls should be adjusted, or the operators should provide some explanation for the deviations. Good communication with the operators will greatly support these efforts.

There are two kinds of errors that show up at this stage, namely time shift errors and magnitude errors. Time shift errors involve the model and field data showing similar patterns, but the times vary, creating a shift. Magnitude errors manifest as different values for properties such as flow and pressure, even after allowing for time shift. Figures 2-3a and 2-3b illustrate these types of discrepancies.

Time shift errors are usually due to errors in demand patterns and control settings. These are difficult to solve but have minimal effect on the usefulness of the model. Magnitude errors are more serious. The modeler needs to determine if they are simply due to operators not following the control rules or if they are revealing some problem (e.g., connectivity, pressure zone boundaries) with the data or model that should have been caught earlier.

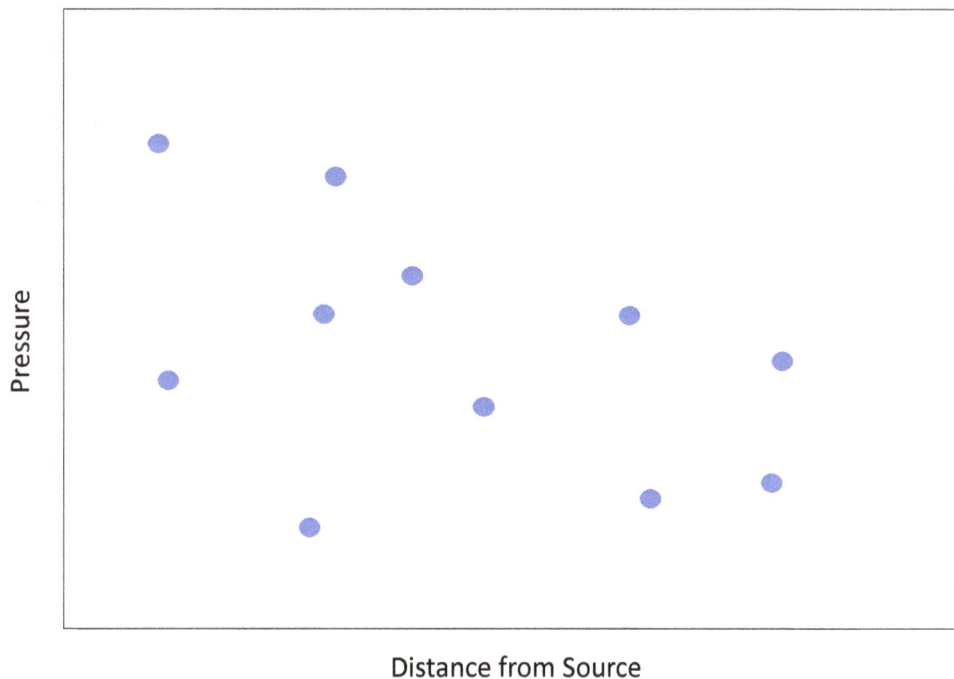

Figure 2-3a Displaying model results in pressure units does not provide a way to determine if values are consistent in working in a hilly area.

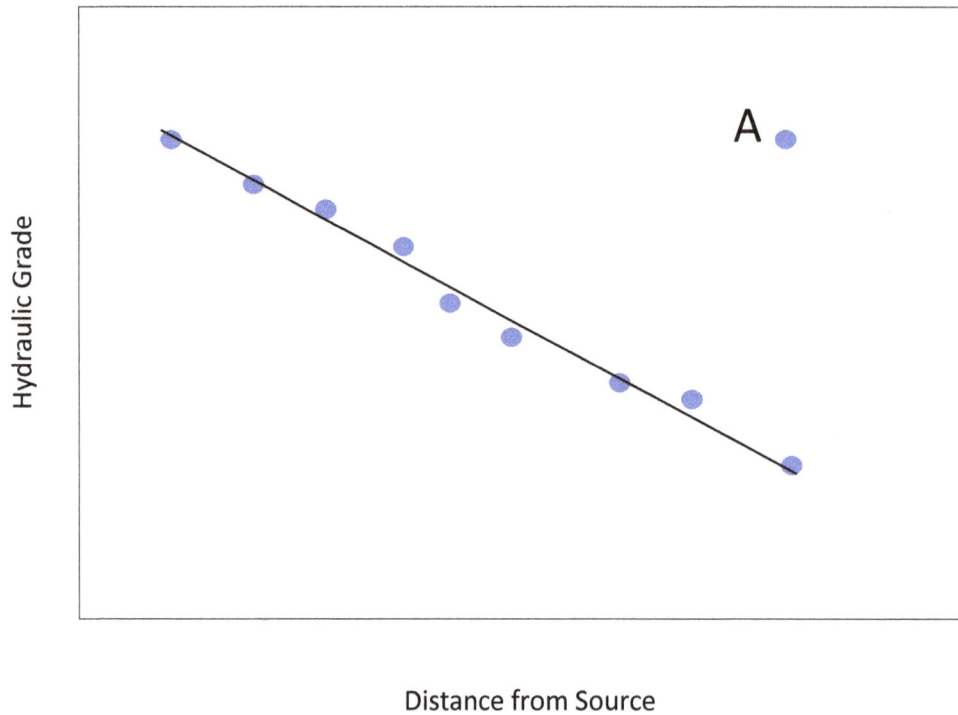

Distance from Source

Source: Courtesy of Tom Walski

Figure 2-3b Plotting field data in HGL units makes it easy to identify inconsistent data such as point A in this figure.

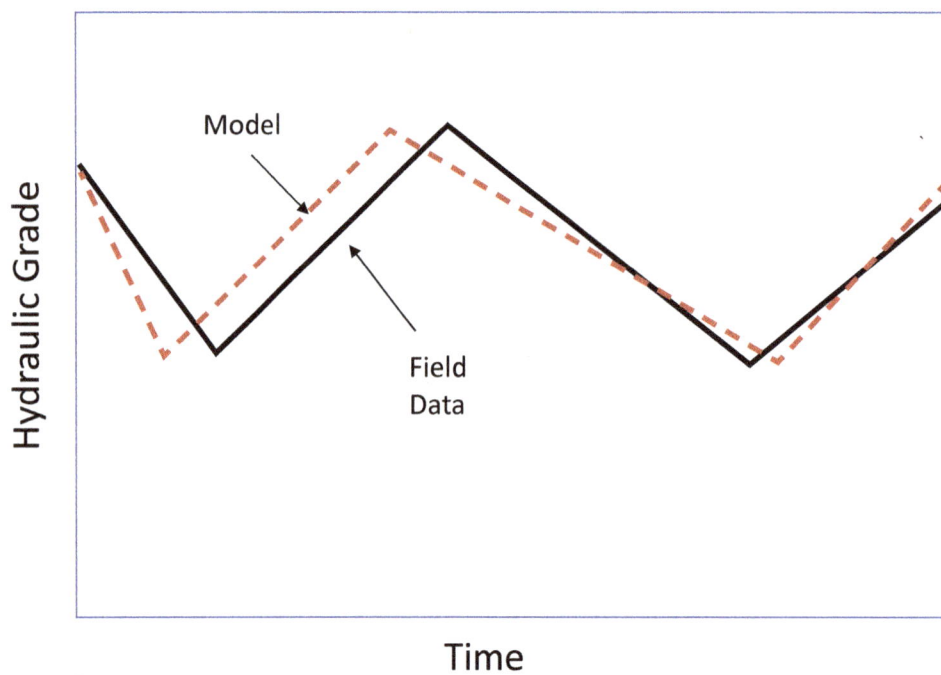

Time

Source: Courtesy of Tom Walski

Figure 2-4a Time shifts in tank water level are often due to small errors in diurnal demand multipliers.

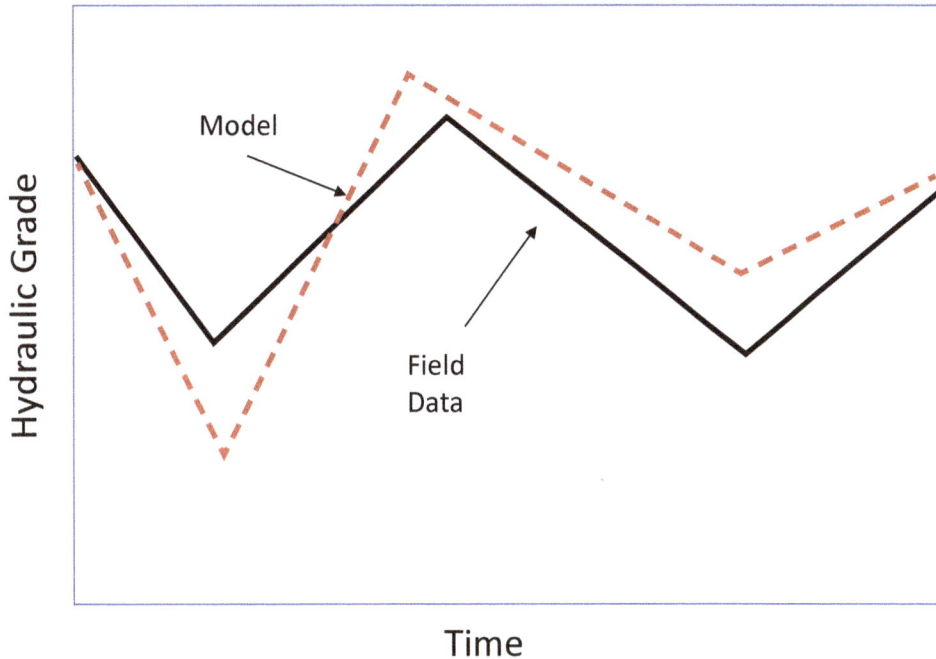

Source: Courtesy of Tom Walski

Figure 2-4b Discrepancies between model and SCADA data that are more than time shift are often due to the rule-based controls in the model not matching operator decisions.

CALIBRATION FOR OTHER PURPOSES

This chapter only addressed calibration for system hydraulics. Models are often used for other purposes such as water quality analysis, real-time operations, or energy use and cost. Before starting any of those types of analyses, it is necessary to have a well-calibrated EPS model that provides a high level of confidence. Otherwise, if water quality or energy results from the model do not match field observations, it becomes very difficult to determine the necessary adjustments.

These models can provide information or insights that can help the hydraulic calibration. Each level of model advancement can be used to verify and adjust previous model parameters to increase accuracy and level of confidence. For example, water quality studies can provide disinfectant decay information or the results of a tracer study which can reveal closed valves or incorrect pipe sizes.

Energy use data can demonstrate where pumps are not working on their curves or rule-based controls are not being followed. For example, control statements may be set up for typical days, but operators may override those control statements.

In some instances, the results of a water quality or energy calibration exercise may reveal shortcomings in the hydraulic model which will need to be corrected.

VALIDATION

There is a range of opinions on the use of the terminologies: "model validation" and "model verification." Model verification is the examination of the numerical technique and computer code to ensure that there are no inherent numerical problems with obtaining a solution; model verification is not a focus of this handbook.

Model validation is the comparison of model results with field data from a time period that was not used during model calibration. In cases where the agreement is poor, it is necessary to understand the contributing factors and recalibrate model, or those portions of the model as necessary, so that it is accurate for both periods.

Some time may pass between the period for which the model was calibrated and when it is applied to solve a problem, prepare a design, or master plan or provide input to operators. Before using the model, validation should be made to ensure that the model still reflects the actual system conditions. For example, for new land development in a specific part of the system, it is advisable to conduct a hydrant flow test at the connection point to check the model's accuracy as calibration data may not have been collected in this specific area.

A related issue is whether a model should be adjusted to match incorrect operational conditions. For example, a model may contain the intended PRV setting or valve status, but in the field, the PRV may have been incorrectly set or a valve incorrectly closed. In these cases, the modeler and operators need to resolve these discrepancies, usually by changing the field conditions rather than changing the model.

MAKING ADJUSTMENTS

Most hydraulic models do not adequately agree with field data when they are first developed or upgraded. The model must be calibrated by identifying why the agreement isn't better and making the proper adjustments to achieve agreement. Faced with a myriad of reasons for discrepancies between a model and field data, modelers can use the procedure described to systematically make adjustments. Limiting the available adjustments at each step should make it easier to identify the source of any discrepancies and correct the model.

REFERENCES

AWWA, 2012, *Computer Model of Water Distribution Systems*, AWWA manual M-32, Denver, Colo.

Ormsbee, L. and Lingireddy, S., 1997, "Calibrating Hydraulic Network Models," *Journal AWWA*, 89(2), p. 42–50, February.

Walski, T.M. 2017, "Procedure for Hydraulic Model Calibration," *Journal AWWA*, 109(6), p. 55–61, June.

CHAPTER 3
Calibration Data
Eric McLeskey

Data collection for model calibration should be carefully planned so the calibrated model provides a reasonable representation of the distribution system that engineers, modelers, and decision-makers can have confidence in. Before data collection begins, the modeling objectives should be defined. Understanding how a utility intends to apply the model should drive the type and amount of data needed to perform a sufficient calibration. It is important to recognize that data collection and model calibration can be an iterative process. As the calibration process unfolds, there may be a need to collect additional data. It is good practice to document data collection by noting appropriate boundary conditions that may impact the data so that the information being collected can be understood in the proper context.

The data available for a water system will vary by utility. While many utilities have Supervisory Control and Data Acquisition (SCADA) systems that can provide insights into how tanks, pumps, and control valves operate on a near-continuous basis, others do not. Field tests can be designed for any system to provide data for model calibration which can supplement, or be used in the absence of, SCADA data.

One of the best ways to understand how a water system operates is to ask the distribution system operators. Discussions with a utility's engineering and operations staff can be insightful and can help the modeler develop an understanding of available data sources. Operational understanding is useful for how water production, storage, and pumping sites operate and the general condition of above ground water infrastructure and data systems as well as areas in the distribution system that may have challenges meeting flow or pressure requirements.

A SCADA specialist should also be included in these discussions to confirm areas where SCADA data is collected and how it can be accessed (i.e., in real-time or through a data historian). The reliability of SCADA data recording equipment (pressure gauges, flow meters, tank floats) can be problematic in some systems so it is important to understand what adjustments or assumptions might be needed to collect adequate data to meet the modeling goals.

Because no two systems are alike, data collection for model calibration is somewhat of a unique exercise. However, there are some general guidelines that can be applied to all systems to assist modelers in collecting useful data that can be used with confidence for hydraulic model calibration.

This chapter explores the data needed to perform a hydraulic model calibration, which is a prerequisite to model use for various hydraulic, water quality, or energy optimization analyses. The data needed to perform water quality and energy modeling is covered in Chapter 6.

DEFINING MODELING OBJECTIVES

Understanding how a hydraulic model is intended to be applied is crucial for developing a good data collection plan. Modeling objectives may even determine if an extended period simulation (EPS) model (a model that represents multiple time periods, typically a 24-hour day) or a steady state (SS) model (a model that represents a single time period) will be developed and calibrated. An EPS model has different calibration data requirements than an SS simulation model does.

Some general questions that can be asked when identifying modeling objectives before developing a model data collection plan include:

- Will the model be used for master planning or capital improvement planning only?
- Will the model be used to estimate available fire flows?
- Will the model be used to perform capacity assurance evaluations for new developments?
- Will the model be used for water quality analysis (water age, chlorine decay, or species modeling)?
- Will the model be applied to solve operational problems?
- Will the model be applied to evaluate energy use and optimization opportunities?

Table 3-1 summarizes several types of modeling analysis that can be completed with an SS model and those that can be completed with an EPS model. It's common to develop an SS model first and then expand it to EPS, but sometimes an SS model is sufficient to complete modeling objectives. For example, it is appropriate to master plan infrastructure and perform fire flow analysis with an SS model. While both tasks can be accomplished with an EPS model, it isn't necessary to develop the additional model detail and perform the additional calibration that an EPS model will require.

All hydraulic models can support distribution system operations to some extent, but generally, EPS models are developed to perform this type of analysis. For example, EPS models are required when evaluating tank fill/drain cycles or tank mixing. Similarly, evaluating pump cycling in response to daily demands

Table 3-1 Data requirements for steady state and EPS hydraulic model calibrations

Modeling Objective	Steady State	EPS
Master planning infrastructure (CIP planning)	Yes	Yes
Fire flow analysis	Yes	Yes
Water quality (age, chlorine decay, source tracing, species modeling)	No	Yes
Energy efficiency	No	Yes
Operational modeling	Yes [1]	Yes
Capacity assurance (estimate impacts from new developments)	Yes	Yes

Note:

(1) Limited, planning-focused analysis is possible; however, evaluating tanks filling/draining/ mixing or pump cycles over time requires an EPS model.

requires an EPS model. However, an SS model can still be used in some situations to support distribution system operational modeling, albeit from more of a planning perspective. For example, an SS hydraulic grade line (HGL) plot can be developed to identify pinch points in the distribution system. Pumps and tanks can be manually adjusted for a single time period to perform sensitivity analysis of the potential impacts of a range of operational conditions. An SS model can also be used to identify system pressure impacts resulting from a pipe break or facility outage, which can inform operations decisions in responding to emergencies.

DATA REQUIREMENTS

Once modeling objectives are established, available data sources and any limitations within data sets should be identified. SS and EPS models have different data requirements. For example, flows and pressures for SS models should be collected for multiple conditions such as average daily, maximum daily, or peak hour demand conditions and for any hydrant flow tests conducted as part of the calibration. For EPS models, continuous flow and pressure readings over a 24-hour or longer period are required.

Similarly, for SS models, data for tank levels and PRV settings are generally used to establish the model boundary conditions and HGL from the sources. In an EPS model, it's preferable to have continuous readings for tank levels, pump station flows, PRV flows, and pressures for calibration.

Table 3-2 A summary of the data requirements for SS and EPS hydraulic model calibrations

Data Requirement	Steady State	EPS
Flow	Yes, multiple conditions (i.e., daily average, maximum day, peak hour, etc.)	Yes, 24-hour
Pressure	Yes, multiple conditions (i.e., daily average, maximum day, peak hour, etc.)	Yes, 24-hour
Tank level	Use as model boundary condition	Yes, 24-hour
PRV pressure	Use as model boundary condition	Yes, 24-hour, if available
PRV flow	No	Recommended, 24-hour, if available
Production data	Yes, simple (supply = demand)	Yes, complex (supply = demand + change in tank volume)
Diurnal demand pattern	No	Yes, by pressure zone
Distribution system pressures (temporary/permanent install)	Recommended	Recommended
Pump curves or design point	Yes	Yes
Pump test to validate Curve	Recommended	Recommended
Pump status/runtime data	Yes	Yes
Reservoir geometry	No, initial tank level sets HGL	Yes

ADDRESSING DATA DEFICIENCIES

When data are not available, reasonable assumptions can be made to approximate the best understanding of how the infrastructure is designed to operate (i.e., the design flow and head for a pump or a PRV set point), or a field test can be conducted with flow meters and gages to collect data for calibration. Sometimes data are collected at field sites (i.e., a booster pump station) in real time but are not recorded by a historian. In these cases, the modeler may need to visit the site, view the screens and gages, and talk with the operators about how the site typically runs in a variety of conditions. While the information gathered in a site visit like this will not typically be used to compare model results with field observations, it can provide the modeler will guidance on how to configure the pump controls, tank levels, pressure settings, etc. associated with these sites to prepare the model for calibration.

HYDRAULIC DATA COLLECTION

The data required for model calibration vary depending on the modeling objective, although there are some common data requirements for steady state, EPS, water quality, and even energy modeling applications to ensure the system is adequately represented established during the model development phase but reevaluated during calibration. This includes

- Physical System Data – system infrastructure (pipes, valves, tanks, PRVs), elevations, pipe roughness
- Operational Data – infrastructure settings (set points, pump curves), operational strategies
- Water Consumption Data – water billing and production data, water demand patterns

The amount and type of data required for calibration will vary depending on the application and types of deficiencies identified. The following sections describe the data required to allow the modeler to adjust the parameters needed to meet their specific modeling objectives.

System Infrastructure

Hydraulic models require an accurate accounting of system components to properly characterize their operation in the calibration and subsequent modeling analysis. This includes data required to build the hydraulic model such as sizes, quantities, volumes, elevations, operating set points, and pump curves. However, the data available may be limited and assumptions or approximations may need to be made. In these cases, the assumptions and approximations should be documented so that they can be re visited during calibration, if required.

Pressure and Flow Measurements

Pressure and flow measurements can be determined using pressure gages and flow meters. SCADA data may also be available to characterize pump station flows, discharge pressure, pump cycling, and/or pump run time. Pressure data can also be collected by reading manual gages, chart recorders, and data loggers.

Field tests can be extremely valuable in understanding system dynamics and nuances that cannot be gained from simply reviewing GIS data or as-built drawings. In developing field tests, it is critical to get input and buy-in from utility operations staff. The operators that manage the distribution system daily know it best and can suggest monitoring locations that will provide the most benefit.

Table 3-3 A summary of the data required for the system components in a water distribution system hydraulic model

Model Element	Represents	Data Required	Potential Source
Junction	Customer location, water demand, change in pipe diameters, intersections	Ground elevation	GIS, USGS digital elevation model/contours, construction drawings
Pipe	Distribution main	Pipe diameter, roughness, length	GIS, construction drawings
Tank	System storage	Bottom elevation, dimensions, water level operating range	Construction drawings, field measurements
Pump	Individual booster pump	Elevation, diameter, pump curve (or design point)	Construction drawings, operations and maintenance manuals, manufacturers pump curves
Control Valve	PRV, PSV, FCV	Elevation, diameter, setting	Construction drawings, operations and maintenance manuals, water system operators

They also can provide insights into equipment issues, problematic infrastructure configurations, changes in the system that are not recorded on drawings or in the GIS, and operational changes that are made to respond to daily or seasonal events.

When field tests are designed, it typically involves placing pressure gages in the distribution system at locations that are far from the water sources. The purpose of this is to estimate the head loss across the system. The best chances for collecting good data from this type of field test is during the summer (or a high-demand period), when head losses are the greatest. A field test conducted during low-demand periods can still be useful, but it is less likely that the results could be used to adjust C-factors or make other determinations associated with the HGL across the system.

It is essential to know the elevations of the pressure gauges in the system for both permanently and temporarily installed gauges. Model elevations should be adjusted accordingly to get an accurate estimate of the HGL. Converting system pressures to HGL is recommended for calibration as it allows the modeler to see if the system is operating in an expected manner and it can also be used to identify potential data errors.

Figure 3-1 shows the operating pressures recorded by four temporary gages that were deployed in a single pressure zone. The average pressures for these four

Sites A - D Pressure

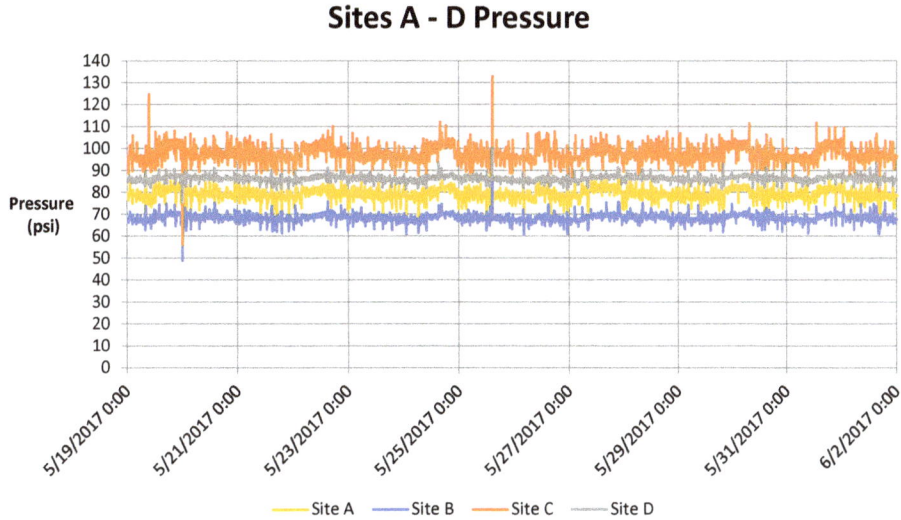

Source: Courtesy of Eric McLeskey, Carollo Engineers

Figure 3-1 Temporary pressure gage pressure readings

Sites A - D Hydraulic Gradeline (HGL)

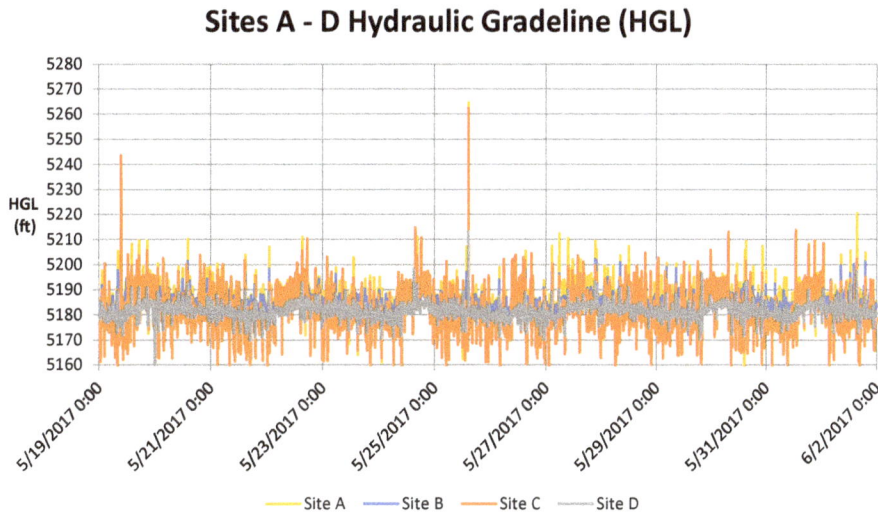

Source: Courtesy of Eric McLeskey, Carollo Engineers

Figure 3-2 Temporary pressure gage HGL values

sites range between 70 (Site B) and 100 psi (Site C). Without knowing the elevations of these sites, it's not possible to know if these individual values or the range of values across the four sites, is reasonable.

Figure 3-2 shows the HGL for the four sites for the same time period. It's clear that these four sites are operating at a similar HGL because the data is nearly coincident. While these data indicate larger ranges of high/low pressures and corresponding changes in HGL across the four sites, the average HGL pattern for this time period shows that these sites appear to be operating on the same grade line.

It is important to understand the accuracy level of the equipment that is providing the pressure and flow data. For example, many pressure gages are accurate to within ± 2 psi, but the equipment may be capable of reporting many more significant digits. Numerical model comparisons with calibration data should be done with this in mind. Therefore, in the case of a transducer that is accurate to within ± 2 psi, report 62 psi versus 62.13835 psi. Additionally, it is important to understand the minimum and maximum pressure recordable by the equipment, as some have maximum limits at 100 psi, while others may extend to 200 psi.

Lastly, collecting large amounts of SCADA data during low-flow conditions is not nearly as useful as collecting data when head loss is significant (e.g., hydrant flow tests) (Walski, 2002).

Hydrant Flows

For smaller pipes (roughly less than 20 in.), hydrant flow tests are helpful for model calibration. Opening a hydrant can significantly increase the velocity in smaller pipes and produce significant head loss. The test involves measuring pressure during normal conditions and again when a nearby hydrant is opened. Steps for running the test are documented in numerous sources (AWWA M17, Insurance Service Office). If hydrant flow test results are used for model calibration, it is important to record not only the hydrant flows and residual pressure but also the boundary conditions of the pressure zone (i.e., tank water level, pump status, and control valve settings).

When conducting fire hydrant flow tests, it is beneficial to use multiple residual hydrant gages. This is useful for detecting where excess head loss occurs and where closed valves are likely. The test procedure is described in Grayman et al. (2006).

Hydrant flow test data provided without the known boundary conditions under which the test was performed has limited use for calibration. However, it could be used as reference data that may uncover an issue with how the model represents the system. For example, if a fire flow test showed 2,000 gpm at 20 psi but the model only provides 600 gpm at 20 psi, there may be an issue with pipe connectivity (closed valve within the model), incomplete data (missing pipes), or incorrect data (pipe modeled as 6-inch main when it is actually 16 in.).

Pipe Roughness Test

While pipe roughness is generally not measured directly, it is very helpful to run pipe roughness tests, where a section of pipe with known material and age is isolated such that the flow, upstream and downstream HGL, diameter, and length are accurately known. This should be done for a variety of pipes in the system but mainly focus on older pipes since they show the greatest variation in roughness. The Darcy-Weisbach pipe roughness or Hazen-Williams C-factor can be estimated from these values. Test procedures are described in Walski et al. (2003). This is especially important when working with systems that contain a great deal of unlined, metal pipes. A table or graph of roughness vs. age can be created for that type of pipe in that system.

Literature values for roughness should not be used because loss of carrying capacity depends heavily on water quality and the use of corrosion inhibitors which can vary from one system to the next (Sharp and Walski). Nevertheless, Lamont (1981) provides some good guidance for initial estimates if a pipe roughness test is not possible.

Pump Performance Test Data

The design pump curve specifies the head added to the system by the pump. Pump performance tests can be used to determine if a pump is operating on its design curve and provide data on the observed head losses within a pump station. The model pump curve will typically include these losses, which will provide less head than the design pump curve. These tests typically involve measuring the flow and pressure for each pump in a pump station under normal operating conditions and by throttling the discharge valve of the pump station. Figure 3-3 shows an example of a pump curve generated from a pump performance test compared to the design pump curve. The field test, in this case, showed that the pump motor was operating below, but within the flow and head range for this pump curve. However, modeling the distribution system with the design pump curve would not provide an accurate representation of the flow or head provided by the pump.

Generally, pump performance test data should be used as model input after they are verified and reviewed. Because pump performance changes over time, it is likely that future performance tests will yield different results.

There may be model scenarios where the performance test data should be used (i.e., calibration) and where the factory curves may be used (i.e., a future scenario that assumes the pump station is reconditioned to its original state).

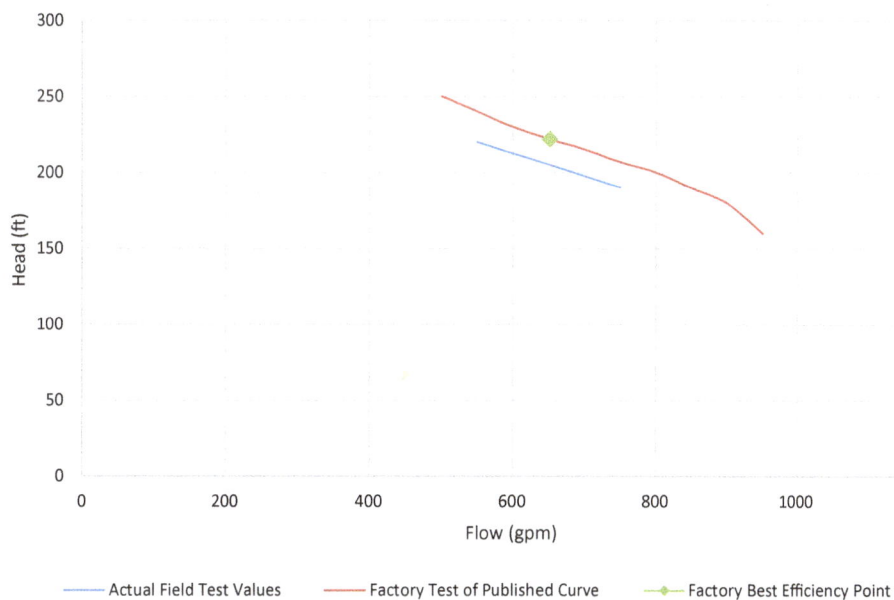

Source: Courtesy of Eric McLeskey, Carollo Engineers

Figure 3-3 Pump flow test results—factory curve versus field test values

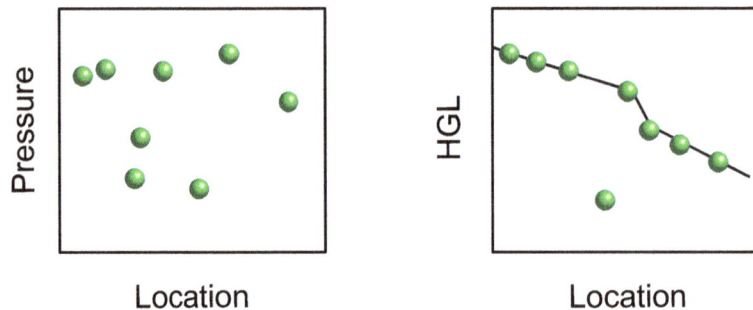

Source: Courtesy of Tom Walski

Figure 3-4 Transmission main system pressure and HGL comparison

Loss of Head Test

A loss of head test can be performed by collecting pressure and flow data along the length of a transmission main. While hydrant flow tests work well for smaller mains, opening a hydrant can have little impact on the velocity in a large transmission main. A better test for this consists of taking pressure and elevation measurements at many points along a transmission main during roughly steady operation (no change in pump status). Plotting the estimated versus modeled HGL as shown in Figure 3-4 can be useful for identifying adjustments in roughness values and identifying closed or partly closed valves.

SCADA Data

SCADA data can be used to determine the boundary conditions for model calibration. When SCADA data are used, it is important to understand the polling interval of the data (time between signals) and whether the data are raw data or have been processed in some way. Data may be transmitted back to the SCADA server every hour, even minute or every second or any other interval.

The modeler needs to understand if the data is instantaneous data corresponding to a time stamp or an average value during the polling interval. As mentioned in Chapter 2, if the polling interval is 17 minutes, a flow reading with time stamp may be 150 gpm at 2:29 but may be the instantaneous value at 2:29 or the average between 2:12 and 2:29. It is important to understand if the SCADA data are raw data or if it has been processed such as by a historian tool.

Ideally, pressure and water level data should be instantaneous while flow data should be average. Data may be raw values from the field or processed when it is saved in a SCADA historian. In some cases, the data may be collected at a high frequency and then "down sampled" and reported as hourly values.

It is also important to understand apparent anomalies in the SCADA data sets. For example, it is not uncommon to see the same value repeated over multiple time intervals, blank values, null values, or values that do not appear to be reasonable. It is important for the modeler to ask questions before proceeding to apply the data. Not doing so, could lead to incorrect assumptions such as that null values on a pump station flow meter mean that the pump was not running. The null value actually may indicate a problem with the SCADA signal, or the data historian and the pump could be running during that time. Additionally, there can be challenges in examining data for individual pumps when SCADA data represent flows and/or pressures downstream of multiple pumps.

SCADA data can also provide some information on system controls. Interviews with operations staff can provide additional insight and may provide understanding of how the system is typically operated on a daily, weekly, monthly, or annual basis. These interviews can also uncover equipment problems (i.e., pumps not cycling correctly) or issues with how SCADA data is received and/or recorded.

Using time-based controls in a model enables the identification of cases where system operators have overridden the rule-based pump control at the time that data were collected. A potential problem in using time-based controls is the length of the polling intervals in some SCADA systems. For example, if a system has a 15-minute polling interval, and the pump is off at 8:00 am and on at 8:15 am, there may not be a way to determine if it was turned on at 8:01 or 8:14 or anywhere in between. A change of slope in the tank level curves may be helpful in determining the exact time of a pump status change. This is especially important for systems with large seasonal variations.

Another consideration in using SCADA data is the need to understand what a value represents. For example, a flow or pressure reading may have the SCADA time stamp 10:42, but that value may be the instantaneous value at 10:42 or the average, minimum, or maximum value between 10:42 and the previous polling time. Another source of error that can be identified at this time is incorrect tank dimensions, but this is seldom a problem in most models.

Data that is typically available in SCADA systems includes:

- Pump flows
- Pump on/off status
- Pump suction pressure
- Pump discharge pressure
- Tank levels
- Tank percent full

Some utilities may also collect data for the following assets but are less common:

- Flows into elevated tanks
- Flow out of elevated tanks
- PRV flow
- PRV upstream pressure
- PRV downstream pressure
- Flows through District Metered Areas (DMAs)
- Control valve status

Demands

Properly characterizing water system demands in a hydraulic model is critical to obtaining a reasonable calibration. The magnitude, timing, and spatial distribution of demands in a hydraulic model are important because they influence how the water sources respond to meet those demands. Water demands can include customers that are:

- Residential (single or multi-family homes)
- Non-residential (commercial, industrial, public facilities)
- Irrigation (turf irrigation)
- Wholesale (private water companies, system interconnects with neighboring utilities)

Each of these customer types generally has different water use habits both daily and seasonally. It is important to understand these differences so that appropriate diurnal patterns (EPS models) or peaking factors (steady-state models) can be developed to represent system water use in a variety of conditions.

Large water users are also important to characterize, and it may be appropriate to analyze these demands separately from the utilities' other demand data to not skew unit water demand (water duty) factors.

WATER BILLING DATA

Water billing data is useful in characterizing a utility's water use and can also be used to geographically reference data in GIS (geocoding) if sufficient location information (i.e., addresses or GPS coordinates) is provided as part of the data set. Generally, water billing data can be used to:

- Review historical water use trends
- Review historical water use by sector (residential, non-residential, irrigation)
- Estimate non-revenue water by comparing with water production data
- Develop a basis for unit water demands

Utility water meter customer billing databases commonly provide monthly or quarterly water use by metered account. It is important to understand that the month the water is billed is not always the month in which the water is used. While it is common for billed water use to be a month after the water was consumed, this time, it will vary from utility to utility based on the metering technology employed, how the meters are read, and how the monthly billing cycles are established. In some systems, there are multiple water meter reading cycles throughout the month, which must also be accounted for.

On an average annual basis, water customer billing data is a generally accepted approach to estimate distribution system water use. This billing data should always be compared to water production data for the same year to determine if there is a reasonable correlation between the water billed and water produced.

Water billing data may not be comprehensive or accurate, especially when' comparing specific accounts (e.g., large users) between multiple years. Accordingly, there are opportunities to either modify some specific demand data or adjust non-revenue water factors within areas as part of the calibration process.

WATER PRODUCTION DATA

Water production records are used in tandem with water billing data to provide a historical understanding of daily, monthly, and annual water demands, which enable calculating peaking factors. The following demand conditions are typically included in hydraulic modeling analysis:

- Average Daily Demand (ADD) – the average daily demand throughout a calendar year.
- Maximum Daily Demand (MDD) also referred to as Peak Daily Demand (PDD) – the highest water production day in a calendar year.
- Maximum Month Demand (MMD) – the average daily demand of the month in a calendar year with the highest water production (often summer months June – August, but not always).
- Peak Hour Demand (PHD) – the highest hourly water demand, which generally occurs on the maximum production day in a calendar year.

These demand conditions are important in hydraulic modeling because they establish the performance criteria that are used to estimate "worst-case conditions" in a distribution system, which can also be important considerations for establishing periods to gather data for calibration. Hydraulic models are applied to evaluate if infrastructure can provide the required flow and pressure without excessive head loss under these conditions. In a steady-state hydraulic model, the average day, maximum month, maximum day, and peak hour scenarios can be developed using peaking factors applied against the average day demand, most commonly for the system, but at times also developed per discrete areas or pressure zones. Peaking factors are a ratio of one demand condition to another. For example, a system with a peak hour demand of 6 mgd and an average day demand of 2 mgd, is said to have a peak hour to average day demand (PH/ADD) factor of 3 (6 mgd/2mgd = 3).

Common peaking factors include:

- Maximum Month to Average Day Demand – MM/ADD
- Maximum Day to Maximum Month – MD/MM
- Peak Hour to Maximum Day – PH/MD
- Peak Hour to Average Day – PH/ADD

Figure 3-5 illustrates the difference between the daily demand, average daily demand, maximum day demand, and maximum month demand.

The peak hour demand commonly occurs on the maximum demand day, but not always. Figure 3-6 illustrates an example that shows the relationship between hourly maximum day and peak hour demand for a system. All systems

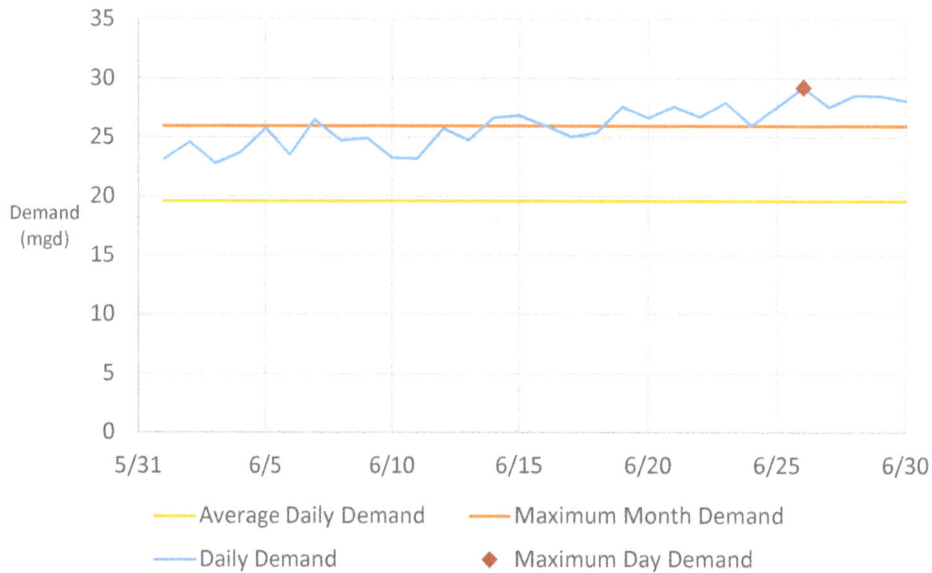

Source: Courtesy of Eric McLeskey, Carollo Engineers

Figure 3-5 Monthly water system demand examples

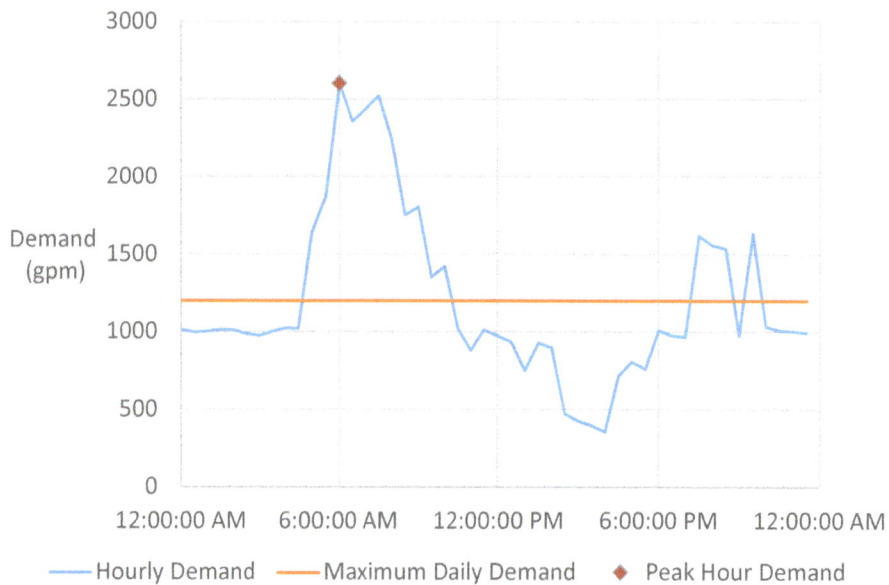

Source: Courtesy of Eric McLeskey, Carollo Engineers

Figure 3-6 Example maximum daily and peak hour demands

will have different demand patterns due to variations in customer composition, the number of residential versus non-residential connections, water use habits such as irrigation schedules and customer use patterns, as well as seasonal weather variations.

WATER SYSTEM MASS BALANCE AND DIURNAL PATTERNS

A water system mass balance is useful to characterize water use within and among pressure zones. In its simplest form, a distribution system mass balance includes:

Water supplies – customer demand – wholesale demand + change in storage (positive or negative).

It is common for pressure zones to have different diurnal demand curves, particularly if land use or development types are significantly different from one pressure zone to another. While AMI allows hourly or daily reads versus monthly, the quantity of data across an entire system can make it difficult to apply it to a model calibration. However, it may be useful in providing insights to customer demand patterns and in developing diurnal curves for the model.

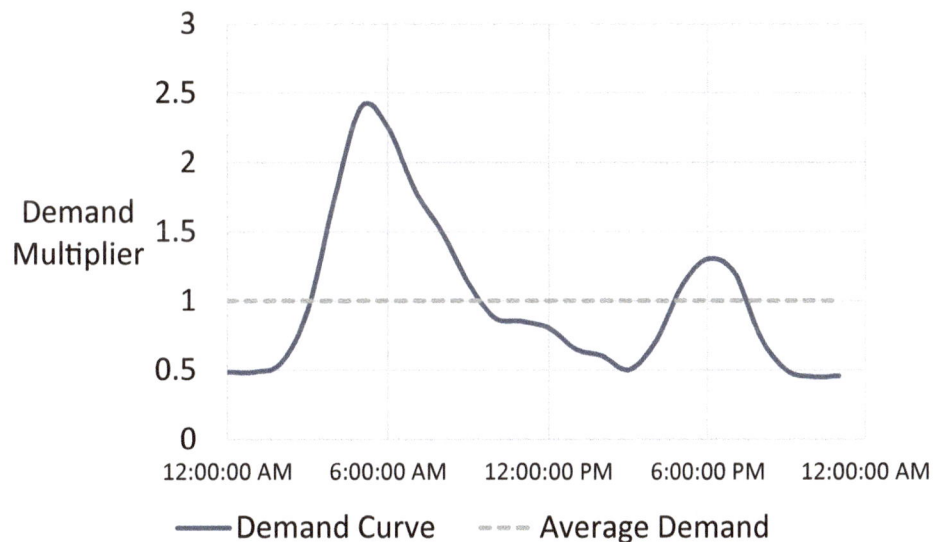

Source: Courtesy of Eric McLeskey, Carollo Engineers

Figure 3-7 Example diurnal demand pattern

Diurnal demand curves can be developed when all the flow inputs and outputs of a pressure zone are known. Figure 3-7 illustrates a diurnal demand pattern with periods of tanks filling or draining to meet customer demands.

DEMAND ADJUSTMENTS

Water demands developed using a water customer billing dataset will generally need to be adjusted for model calibration using appropriate factors developed by understanding the system water production and the system operations as determined from the system mass balance. These adjustments should also account for wholesale customers that take water from the system or any connections that result in flows into the system (i.e., if the utility purchases bulk water from an adjacent provider.)

Demand adjustments for model calibration should be reasonable when compared to actual seasonal demand changes. For example, if a system typically has 40% higher demands in the summer than on an average annual basis, a scaling factor of 1.4 applied to the annual average (water billing) data would be an appropriate starting point for a calibration day that is in July. This is assuming the demand adjustment is verified by reviewing the water production and operating data for the calibration day.

Demands should be adjusted to match the calibration day utilizing production data. New diurnal curves and daily demand should be adjusted to ensure demands in the model represent the demand conditions during calibration system wide. If possible diurnal curves should be developed for smaller monitored areas within the distribution system where data are available.

DATA COLLECTION PITFALLS

It is important to be aware of common pitfalls that can introduce error in the calibration data. The purpose of this section is to identify things that can cause problems in the data analysis and steps that can be taken to avoid these issues.

The following outlines common problems in collecting data for model calibration:

Uncalibrated Gages: Equipment that is not calibrated introduces the potential for error that likely cannot be resolved without re-testing. All those handling the instrumentation should be trained in how to handle, install, and take readings. A post-field test calibration can be performed on equipment, if necessary, to determine if the equipment is operating correctly.

Pressure Recorded at Unknown or Incorrect Elevation: USGS maps are not sufficient to estimate elevations for instrumentation. GPS survey data should be collected (if not already available). Open data such as USGS and Google Earth can be used as reference data (i.e., may be used to support a decision to deploy a survey) but are not accurate enough to be relied on for model calibration. The level of accuracy needed to measure elevation for instrumentation will depend on the complexity of terrain and elevation ranges within the distribution system.

Transcription Errors: May result from instrumentation not being clearly marked so when it is collected from the field it is not clear where each pressure gage or flow meter was recording data. GPS-enabled applications or GIS maps for field tests and tables with locations matched to equipment IDs reduce chances for mixing up data collection locations.

No System Data Is Available: If system SCADA data are not available, a field test will need to be designed to collect system flows and pressures as well as system boundary conditions (tank levels, number of pumps running, flows in and out of pressure zones or DMAs).

Historical Pressure Gage or Flow Test Data: If historical pressures or flows are provided without the system boundary conditions known for the period that the data are provided, their value in model calibration will be limited. This type of data can be useful as a reference, particularly for areas of the system that may have habitual problems.

Field Test During Low-Demand Period: Sometimes conditions (e.g., project schedule) drive a water system field test to be performed during a time of year or period with low system demands. Ideally, a water system field test will be conducted during a period when demands are high and the system is at its most stressed condition. Data collected during high-demand periods is more useful than data collected during low-demand periods because the increased system head loss during high demands increases the chance that uncertainty from instrumentation error in measuring head loss is minimized. If a water system field test must be completed during a low-demand period, the engineer must be aware that there may be limitations in applying the model during high-demand conditions (i.e., peak hour or fire flow conditions). There may be water quality considerations to consider that may drive when and how field tests can be completed. For example, conducting tests overnight because of the potential to create reverse flow conditions and increase water quality complaints.

Pump Characteristics are Unknown: Pump curves are a good starting point for understanding system pumping capacity. However, as pumps wear over time, they gradually shift and operate off their curves. Pump tests can validate the operating points of individual pumps, which provide a more accurate picture of the HGL leaving a pump station and, in turn, reduce uncertainty in model calibration.

SCADA Data Uncertainty: Generally, SCADA systems can provide an average or sampled value for each point. It is important to clarify with the utility providing the data which values are needed and confirm that the appropriate data was provided. In developing diurnal flow patterns, sampled values are generally better for tanks with a longer time step because tank volumes gradually change over time and this approach reduces the error in estimating flow in/out of tanks. In developing diurnal flow patterns, average values are generally better for pumps as their on/off cycles result in more instantaneous change. Average values account for this and prevent over predicting flow volumes from pump stations throughout the day.

SCADA Data Are Incomplete: If it is not possible to develop diurnal flow patterns and demands by pressure zone due to a lack of data, it may be necessary to group pressure zones in a way that the data can support model calibration.

SCADA Sensor Locations Not Ideal: Knowing that a system has SCADA is not enough to determine if that data can be used for a model calibration. Sensor locations should be confirmed with the utility to make certain they are located appropriately and that they are operating as expected. If tank level sensors are located a significant distance from the site, head loss in the pipe to/from the tank may need to be accounted for in the tank level reading for that site.

Incomplete Understanding of System Operations: Operators are the best source for understanding how a system operates. However, relying on "typical" set points, design points, for model calibration can be problematic and can miss changes operators make on a daily or seasonal basis.

REFERENCES

AWWA, (2016) "Installation, Field Testing and Maintenance of Fire Hydrants," AWWA Manual M-17, Denver, Colo.

Insurance Service Office "Fire Flow Tests," Atlanta, Ga.

Grayman, W. et al. (2006) "Calibrating Distribution System Models with Fire-Flow Tests," *AWWA Opflow*, April.

Lamont, P. (1981) "Common Pipe Flow Formulas Compared with the Theory of Roughness," *Journal AWWA*, Vol. 73, No. 5, p. 274, May.

Sharp, W., and Walski, T. (1988) "Predicting Roughness in Unlined Metal Pipes," *Journal AWWA*, Vol. 80, No. 11, p. 34, November.

Walski, T. (2000) "Model Calibration Data: The good, the bad and the useless," *Journal AWWA*, Vol. 92, No. 1, p. 94, January.

Walski, T., et al. (2003) *Advanced Water Distribution Modeling and Management*, Bentley Systems, Exton, Pa.

CHAPTER 4

Detailed Calibration Process

Matt Huang, Tom Walski, Aurelie Nabonnand

This chapter provides a hydraulic model calibration framework and methodology for water distribution systems. As discussed in Chapter 2, hydraulic water distribution system models can be used in a wide variety of applications from supporting design, planning, to solving operational problems. Since these tasks may result in engineering decisions involving significant investments, it is critical that hydraulic models provide an acceptable representation of the real water system and that modelers and decision-makers have confidence in the model predictions. There are different levels of calibration based on model use: planning level vs. water quality vs. design. The goal in calibration is to reduce uncertainty in model parameters to a level such that the accuracy of the model is commensurate with the type of decisions that will be made based on model predictions.

To confirm that hydraulic models accurately represent the real water system, it is customary to measure various system metrics (e.g., pressure, flow, or storage tank water levels) during field testing and then compare the field results to model predictions. Field data collection is discussed in Chapter 3. However, models can be calibrated only for a specified subset of system conditions. For each model, it needs to be explicitly stated for which conditions the model was calibrated.

If the model adequately predicts the field measurements under a range of conditions over different periods of time, the model can be considered as calibrated. If there are significant differences between the measured and modeled data, further calibration is needed. There are no general standards for defining what is adequate or what a significant difference is. However, it is recognized that the level of calibration required will depend on the intended use of the model. A greater degree of calibration is typically required for models that are used for detailed analysis, such as design and water quality predictions, than for models used for more general planning purposes (e.g., master planning). Each model is best suited for use under the conditions for which it was calibrated. Accordingly, users should keep the following examples in mind:

■ A model calibrated in one season may not be applicable for use in others.

■ If a model is not calibrated for fire flows, fire flow representation may be inaccurate.

■ If a model is calibrated for a steady-state condition, it may not simulate the system well over time.

Calibration should be performed with an understanding of the eventual use of the model. For example, if the model will be used for a fire flow analysis, calibration should focus on the use of hydrant flow tests. If it is to be used for establishing pump controls, pump testing and EPS (extended period simulation) runs should be emphasized.

It is important to define the calibration period, as understanding the conditions the model will be calibrated to is critical.

Questions for the modeler to ask include the following:

1. What demand and operational conditions are the model calibrated to? Is this condition applicable for the system evaluations to be performed?

2. If the model is calibrated for EPS, what is the time period for the calibration? One day? multiple days or weeks?

3. If hydrant tests are required, can the tests be performed in the summer to increase stress on the system?

Case Study 1: Non-Pressurized Pipe

A utility had a large diameter transmission pipeline that, no matter what they did, they could not effectively model the hydraulics of the large diameter pipeline: the pressure in the model was always higher than the pressure in the field. Eventually it was discovered that the pipeline had downstream control and the pipe was flowing by gravity in portions of the pipeline. The systems and calibration analyses presented here require that the hydraulic grade is controlled upstream, and pipelines are pressurized with full flow conditions. When a system has pipelines flowing by gravity in a water system, often a different software must be used to model open-channel flow conditions. Most, but not all, water distribution modeling software can also be used for full pipe situations.

Lesson Learned: Understand the system hydraulics and select appropriate software for the conditions faced.

CALIBRATION PHILOSOPHY AND APPROACH

The number of potential parameters to consider during calibration can be overwhelming. This chapter is structured to introduce the most common parameters to consider first, while the rest of the sections discuss potentially less common parameters that would need to be considered if the model is still not adequately calibrated. Each parameter will evaluate the impact on model results and in which case they should be adjusted. This chapter describes a logical approach to model calibration and guides modelers on the parameters to consider and adjust, instead of randomly adjusting parameters. After setting up the model, the first step in the calibration process is to identify reasons why the hydraulic model does not agree with field results. This step is critical and typically more difficult than adjusting parameters.

Calibration Setup

The main task in calibration is a comparison of model results with data collected in the field. Based on the field data available, a specific time must be selected as the calibration period, in which the field data will be compared to the model data. Events that affect system performance in ways that are challenging to represent (e.g., water main breaks, large sporting events) may not be ideal to use during the period for calibration. Once the appropriate period is selected, the model needs to be set up in such a way that the boundary conditions match field conditions. These boundary conditions most often include:

- Demands – Model demands should reflect actual demands at the time of calibration, which can include wholesale demands, as applicable.
- Tank levels – Tank or reservoir levels should be set so that the tank level matches field conditions. For an EPS model run, this level would be the initial level of the tank or reservoir at the beginning of the simulation.
- Pump and valve status – Pumps and control valves should have the same on/off status in the model as in the field. EPS model runs should match the same status at the beginning of the simulation.

Once the boundary conditions have been represented, the model can be run, and comparison between model results and field data can begin.

Identify Potential Reasons for Differences Between Modeled and Measured Values

The process of calibration requires a systematic approach to identifying potential sources of error or causes for differences between modeled and measured values across multiple parameters and inputs and iteratively making adjustments to reduce those differences. This process is represented in Figure 4-1. Differences in results during the calibration process can come from the following sources. Refer to Chapter 2 for a detailed list. This section focuses on the most common errors.

- Physical properties and network:
  - Connectivity—pipes inaccurately represented as connected or disconnected.

- Pipe properties—inaccurate pipe diameter or material. Older pipes may have a reported diameter based on a default value, operator, or field crew recollection, which may require verification. Additionally, some pipes may have been lined or replaced but not updated in GIS or other records.
- Tank properties—inaccurate tank data such as dimension/shape, diameter, and finished floor and overflow levels
- Zone boundaries—inaccurate zone boundaries can be the product of adjustments made by field crews for temporary needs, which are not adequately recorded or relayed to engineering staff or readjusted in a timely manner.
- Elevations—elevation data at facilities may not be accurate. Model junctions are often assumed to be at ground level. Pumps and valves may be above or below ground level; therefore, adjusting the elevation of pumps and valves to the elevation of the pressure gage or transducer can improve agreement of the data sources
- Demands:
  - Non-revenue water—often the difference between production and metered demand data are equally spread across the entire system. If some areas of the system have higher than average leakage rates, emitters or extra demand may need to be allocated to those areas.
  - Operational settings and field testing:
  - Worn pumps—pump curves are lower than original manufacturer curve

Supervisory Control and Data Acquisition (SCADA) data—sometimes SCADA data can have gaps or are inaccurate.

Adjust Parameters to Minimize Differences Between Modeled and Measured Values

As discussed in Chapter 2, there are many reasons for which a model may not be acceptably calibrated. The key step here is understanding WHY the model doesn't agree with the field data. Once the modeler understands why, the appropriate adjustments can be made.

Adjusting the model is an iterative process. It is typically best to perform comparisons between the model and field data starting from simple scenarios to more complex scenarios. This typically means starting with low-flow steady-state periods, followed by high-flow steady-state periods, then to EPS so that the number of parameters to adjust is manageable. However, these steps can be adjusted based on the intended use of the model, available data, and accuracy of the existing model.

Usually, the first parameters to adjust are pump curves and on/off set points (set points don't become a consideration until use of EPS models). For the initial steps, the user should know the pump status, which may require some investigation if status is not available through SCADA. Next parameters to adjust would be valve settings and controls. Control valve flows, if present, should be checked concurrently as these values can be influenced by those from pumps. There may be instances where no level of reasonable adjustment will help adequately match model results with field data. In this case, the field data may be inaccurate, other unknowns (e.g., closed valves) have not been identified, or the modeler may be adjusting the wrong parameter. Therefore, adjustments under different conditions or through an EPS simulation can help identify the source of problems, including controls and patterns.

Under high-flow periods, the head loss can be increased to reveal other issues not identified under low-flow periods. If adjusting pipe roughness within reason improves calibration, then it can be attempted. In general, pipe roughness should be adjusted in groups according to age and material and potentially location. If a system is supplied by multiple sources with different water quality, there may be pockets where C-factors of the same material and age have different pipe roughness values. Without supporting evidence on field data, adjusting single pipeline roughness is not preferred, even if it leads to closer calibration. Details and case studies are represented in the sections below.

Once the model is calibrated such that the system hydraulics have been adequately represented, the modeler can potentially perform additional calibrations for more advanced applications such as real-time, water quality, and energy evaluations which are covered in subsequent chapters.

Reasoning for parameter adjustments. The modeler must know which parameters to adjust to minimize differences between model results and field values. Some examples of common differences and potential considerations are listed below.

- Excessive head loss within the network
  - Connectivity between pipes may be inaccurate. Review of pipe diameters and potentially disconnected pipes in the area of high head loss would be warranted. Similarly, valves represented as closed that are normally open may have a noticeable effect.
- Pump suction and discharge head:
  - Discharge and suction head difference can be plotted against pump flow in a scatter plot, and the pump curve used in the model can be superimposed on the field data. Field data points falling below the manufacturer curve would seem to indicate a worn pump. A "worn" pump curve can be generated and used in the model to get a better calibration as discussed in Chapter 3.
- Tank levels:
  - Demand supplied from the tank may be too low in the model. This could also indicate excessive leakage downstream of the tank.

■ Tank geometry may not be correctly represented.

■ Upstream pumps may be worn and thus pump curves may have to be adjusted to compensate for a change in performance.

■ Control valve pressures and flow rates.

■ Many control valves do not have field data available. Temporary flow and pressure data logging may be helpful in these areas since flows through control valves can significantly change model results. At times, control valves can stray from their set points unless routinely inspected, maintained, and readjusted. Therefore, adjustment of these set points within the model can be justified, especially after consultation with operations staff. If the model has the "correct" control valve setting but the control valve has drifted, the modeler and operator need to discuss whether the model should be adjusted to match the "as found" condition or the valve should be adjusted to what it was supposed to be. This is also the case when an isolation valve is found not in its correct position. Engineering judgment should be made on the level of effort taken in calibration, especially when reconciling issues tied to controls. For example, a very small area supplied by one or two PRVs (pressure reducing valves) with no outlet to a lower gradient zone may not likely warrant pressure monitoring and a detailed level of calibration, and documented PRV settings can be applied without further evaluation unless projects or issues in that area are a concern.

Case Study 2: Ice Cube

In April in a northern US town, a fire flow test was run near the edge of the system in an industrial park. The pressure dropped more than it should have during the test especially considering that there was a ground tank on a nearby hill. It was suspected that there could be a closed valve between the tank and the flow test. But it was getting late in the day. What else could it be?

Before looking for closed valves, the crew decided to check the tank based on telemetry data from the treatment plant. Circular pen and ink charts were being used. The chart for the previous day showed a perfectly circular trace, same for the days prior. The tank water level didn't appear to be changing. The crew climbed on the tank and opened the hatch to find a two-million-gallon ice cube on the hill. The altitude valve apparently froze that winter, and no one had noticed or checked the data.

Lesson learned: Think about what your data are telling you. Operators can often troubleshoot issues through review of telemetry/SCADA data.

Case Study 3: Sawtooth HGL (Hydraulic Grade Line)

A young consulting engineer had received some pressure data from a water system for a small boosted pressure zone with a tank on the hill. For any combination of tank level and pump status, some of the pressures would match the model and some would be far off. What could be done?

The senior engineer converted the pressure data into HGL values that demonstrated a saw-tooth profile with no consistent slope. When asked about when the pressure data were collected, the young engineer said one value was obtained in the morning in May, another in an afternoon in July. The steady-state model was being used to represent conditions for different time periods and not a snapshot in time of the system. The field data did not correspond to a snapshot but rather different times and boundary conditions.

Lesson learned: Know when the data were collected and the boundary conditions (especially pump status and tank levels) associated with those times.

Sensitivity analyses. A sensitivity analysis can help inform which model variables make the biggest impact on results when changed. A user can modify a certain parameter over a reasonable range and observe its impact on model results. If changes can bring the model into better agreement at some field points without making it worse at others, that adjustment may be appropriate. If the effect is negligible or it makes the agreement worse, it is not likely the appropriate parameter to adjust. However, simply because a model is sensitive to a change in a parameter does not necessarily mean that the parameter is the correct one to adjust to achieve calibration; therefore, it is important that the results from these sensitivity analyses be documented and revisited as other adjustments are performed during the calibration process.

CONDITIONS AND CONSIDERATIONS FOR CALIBRATION

There are a number of parameters that can be modified to calibrate a model. The key to a good model calibration is making modifications so that the model more closely reflects field conditions, and the modifications made to the model are reasonable. This section describes the parameters that could be modified during model calibration. Adjustments generally fall into three categories:

1. Adjustments during steady state with normal low-flow conditions

2. Adjustments during steady state with periods of high head loss.

3. Adjustments during EPS conditions

The considerations for calibration with adjustments under these different conditions are shown in Table 4-1. While the considerations made for adjustment are grouped per condition, they may be applicable under other conditions. A common example is network connectivity, where pipes connect in the physical system, but are not represented as such within the model. This scenario is often identified as an issue during high-flow conditions based on noticeable head loss or pressure reduction; however, the same condition may be apparent under normal or EPS conditions as well.

Chapter 2 presents the general approach toward calibration under these conditions.

Considerations for Calibration During Steady-State Normal-Flow Conditions

Various model parameters can be adjusted during a steady-state simulation with normal low-flow conditions.

Tank water levels. For calibration purposes, levels for tanks should match the level of the SCADA data. When comparing hydraulic grade or pressures at nodes, the tank level in the model run should be the same as the tank level in the real system when those pressures are read. However, because tank level serves as a boundary condition during a steady-state simulation, it is important to verify this value is accurate. After calibration, an average or typical starting level can be used for each tank when applying the model to other conditions.

Table 4-1 Conditions and considerations for calibration adjustments

Conditions	Considerations for Calibration
Steady State: Normal flow	▪ Tank water levels
	▪ Elevation of nodes
	▪ Elevation of pressure sensors
	▪ Pump curves, wear, and variable speed
	▪ Pressure zone boundaries
	▪ Control valve setpoints
Steady State: High flow	▪ Pipe roughness
	▪ Closed isolation valves
	▪ Spatial demand allocation
	▪ Connectivity
	▪ Pipe size
	▪ Excessive skeletonization

(continued)

Table 4-1 Conditions and considerations for calibration adjustments (*continued*)

Conditions	Considerations for Calibration	
Extended Period Simulation	▦	Pump/Valve controls
	▦	Manual operation
	▦	Demand patterns
	▦	Tank dimensions

Elevation of nodes. Elevation data within the network may be available through various sources. These may be recorded by different crews and at different time periods using different equipment or available sources. Sometimes vertical datums change as well; therefore, elevation data used in modeling should be corrected to a common datum before use. Elevations at junctions within the network are commonly ground elevation, while elevations for tanks, pumps, and valves should be the elevation of the pressure transducer or elevation above the finished floor.

Trick: Gauge vs. Absolute Pressure

When collecting pressure information in the field, be careful that pressure loggers can measure in gauge pressure or absolute pressure. The differences between the two are significant, and one must be careful to confirm that the pressure loggers are measuring in gauge pressure. If the loggers are measuring in absolute pressure, the results must be adjusted before they can be used for model calibration.

Lesson learned: Check what is being measured by the pressure logger before installation.

Trick: Different Pressure Reference Points

The elevation of the pressure sensor may not be the same as the elevation of the corresponding model node. For example, the model node may be at elevation 400 ft, but the hydrant outlet where the pressure gage was connected may be 410 ft. If the pressure gage read 60 psi (139 ft), the hydraulic grade is really 549 ft, not 539 ft which would be the value if node elevation is used. For this reason, comparisons between the model and field pressure data should preferably be made in hydraulic grade units, not pressure units. In addition, comparisons made with hydraulic grade unit make it easier to identify outlier data where a pressure or elevation may be incorrect.

The model based on ground elevation should report 58.7 psi but if you read pressure at different elevations, you may be trying to match 61.8 or 50.5 psi.

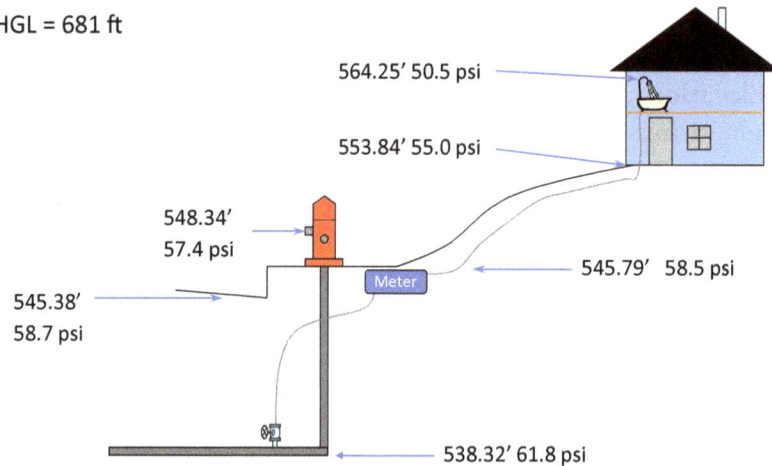

Correct HGL = 681 ft

564.25' 50.5 psi

553.84' 55.0 psi

548.34'
57.4 psi

Meter

545.79' 58.5 psi

545.38'
58.7 psi

538.32' 61.8 psi

Source: Courtesy of Tom Walski

Figure 4-1 Different pressure reference points

Lesson learned: It is often much easier to work with hydraulic grade when comparing values from different locations.

Models typically have elevations assigned to model nodes, as well as to facilities such as tanks, pumps, and valves. It is especially important to make sure that elevations at facility nodes are represented correctly; use of record drawings, where available, is recommended to confirm that key assets within facilities are assigned the correct elevations.

Elevations could be assigned at the physical pipe elevation or the ground elevation in water systems. Both are acceptable, but due to the greater use of GIS to assign elevations into the model from contours, digital elevation models (DEMs), or some other electronic source, ground elevations are more typically used. It is most important to be consistent and understand the accuracy of the elevation data.

Vertical datums may have changed over the years; it is important to note what vertical datum is used, especially for record drawings at facilities. The two most common vertical datums in the United States are National Geodetic Vertical Datum of 1929 (NGVD29) and the North American Vertical Datum of 1988 (NAVD88); some utilities may have older projects on the older datum and newer projects on the newer datum.

When using electronic elevation sources such as contours or DEMs, understanding the accuracy of the elevation source is important. For specific locations, elevations may need to be modified to better reflect field conditions, such as hillside locations.

Case Study: Hydrant Elevations

A static calibration was conducted for a fire flow test located in a relatively new subdivision. The model recorded both that static and residual pressures to be approximately 10 psi lower than the field data. When the static pressure doesn't match between field and model data, hydrant elevations are often the source of the error. USGS contours were used as the source of elevations in this model. In this case, it was discovered that there was significant grading of the site when this subdivision was constructed; over 20 feet of material was removed, and the elevation of the site was over 20 ft lower in reality compared to the contours that were used to determine the elevation for the model. The elevation data was several decades old and did not appropriately reflect the current conditions.

Lesson learned: Elevation sources do not always reflect current conditions.

Elevation of pressure sensors. Tanks may use a pressure transducer or some other means of recording and/or calculating tank level. Periodically these electronic readings should be checked by a manual read to make sure the sensor is functioning correctly.

Case Study: Level Indicator Inaccuracies

Could not match tank level with SCADA, client went back in the field to perform manual measurements and compare to SCADA, SCADA level was off by 4 feet. Inaccuracy in water levels usually show up as large discrepancies between field and model hydraulic grades near the tank with smaller errors as one moves away from the tank.

Lesson learned: SCADA information is not always reliable.

Pump curves. Typically, the pump performance curve provided by the pump manufacturer is entered into a hydraulic model. Occasionally, when that information is not available, pump curves are represented in the model as a single design point or using a three-point curve. Often the operating point of a pump will not match the manufacturer's curve due to pump wear or variable pump speed. One way to check the operating point(s) of a pump, when SCADA data is available, is to calculate the pump head from suction and discharge pressure and compare the pump characteristics (head and flow) with the pump curve, adjusting the pump curve as necessary. If a pump is operating at a point not on its curve, then there may be a mechanical or hydraulic problem with the pump. If it is operating on its curve but at the wrong point, then the issue may be associated with the network upstream or downstream of the pump. For instance, a valve downstream of a pump may be partially closed or throttled depending on the demands downstream. If SCADA data record flow and pressure for multiple pumps (e.g., downstream of a discharge manifold), then it will be important to analyze data representing conditions when each pump is on individually.

Pump wear. Pumps wear over time and operate like a pump with a smaller impeller. This can be adjusted by manually adjusting the pump curve within the model.

Variable speed pumps. Some pumps have variable speed drives (VSP) to modulate the pump speed. In some cases, a VSP is modeled using a pump in series with a pressure reducing or flow control valve; however, the associated energy use and costs cannot be developed if the pump is modeled in this way. In some modeling programs, VSPs can be accurately modeled based on a pressure setting. The speed can also be set using a pump speed setting if the speed is constant throughout the simulation. Some model software packages also include a VSP feature. When collecting calibration data for a pump, it is important to not only measure flow and head but also the speed corresponding to those hydraulic readings. At a given speed, the flow and head points can be overlaid on the pump curves corrected for speed to see if they match.

Case Study: Cavitation to Death

The flow predicted for an in-line booster pump serving a remote portion of a distribution system was much higher than measured. The mystery continued until operators decided to check the pump performance. It was much poorer than the curve for when the pump was new. The impeller was removed and was found to have been severely damaged by cavitation. The new impeller performed as expected in the model.

Lesson learned: Centrifugal pumps are much better at pushing water than pulling water and in remote portions of a system excessive head loss may lower the available net positive suction head such that cavitation occurs. Pump performance should be checked as pumps don't always behave as reported in their original pump curves.

Case Study: Lead and Lag

A pump station fed a small pressure zone with an elevated tank. The station had two pumps and the lead and lag pump switched on each cycle. On some cycles, the pump data matched the model while on others, the pump put out much less flow. What caused that?

Examining the SCADA data from the pump showed that when one pump was running, the performance was on the pump head curve while the other pump was running well below its curve. The problem was not with the model but was a mechanical problem with the second pump. When repaired, the pump performance matched the model.

Lesson learned: Not all pumps run on their pump curves.

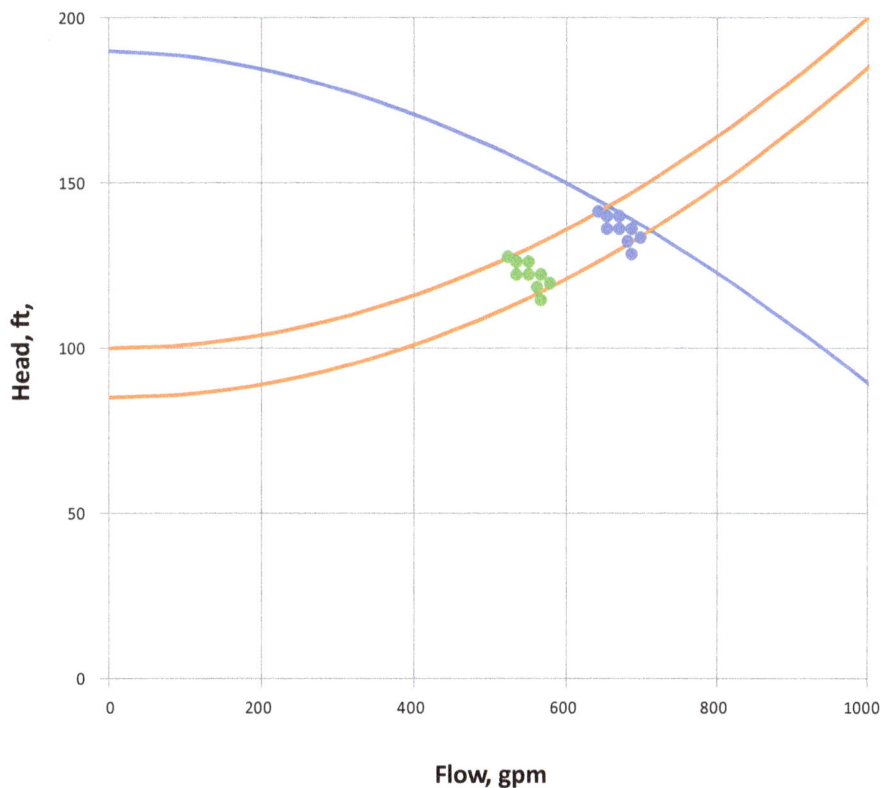

Source: Courtesy of Tom Walski

Figure 4-2　Illustration of a pump not running on the pump curve

Pressure zone boundaries. Pressure zone boundaries separate the water system into hydraulic zones by closed isolation valves. These can be modeled by closed pipes, closed isolation valves, or closed throttle control valves (TCVs). In model development or calibration, one sometimes finds that the zones are not separated at the correct location in the model. Often, during model calibration, it is the case that not all isolation valves are closed in the model between pressure zones, allowing water to flow between two zones. Occasionally, one will find that field conditions are not as expected, that there are open or closed valves in the field. If the hydraulic grade slopes steeply in the model compared with the field data near a pressure zone boundary, then it is likely that the model has an open connection between the zones that is really closed. Conversely, if the field data show a gradient near the boundary that the model doesn't, then it is likely that there is an open connection in the field. Measuring static pressures along the pressure zone boundary can reveal the problem. If water quality data are collected in two supposedly hydraulically isolated zones, the samples may also suggest open boundary valves if values in the zone with lower hydraulic grade reflect those of the higher zone.

Case Study: Hydrant Test – Field Valves

A person was calibrating a city's model to data from a fire flow test. Railroad tracks bisected the system and were laid several years before the model was initially calibrated. The water sources were all located on one side of the railroad tracks; the city had approximately half of its demand on each side of the railroad tracks. When the model was calibrated, pressures on the side of the city with the sources calibrated well. Pressures on the side of the city across the railroad tracks all had lower pressure in the field compared to the model, especially residual pressures. After some research, it was found that, when the railroad tracks were constructed several years prior, an isolation valve was closed at one of the railroad crossings. However, the valve was never reopened at the end of construction, and therefore, a major isolation valve was closed in the field while the model did not reflect that closure. With the valve closed, model calibration improved, and the City was able to reopen the valve to rectify the hydraulic bottleneck. Similarly, the process of model calibration also identified isolation valves that were mistakenly open or partially open between pressure zones, allowing flow through a location thought to be hydraulically isolated. A closed isolation valve was identified based on field data indicating excessive head loss, especially during high-flow periods.

Lesson learned: Isolation valves that one expects to be open might actually be closed in the field. Isolation valves that one expects to be closed might actually be open in the field.

PRV and other Control Valves Settings

Case Study: How Open Is Open?

A neighborhood was supplied by a 12-in. pipe that was itself supplied by a 24-in. and 8-in. pipe. While field static pressures matched the model, during a fire flow test, the pressure dropped much more than expected. The pressure did not drop on a second residual hydrant on the 24-in. pipe. The logical conclusion was that the valve connecting the 12-in. and 24.-in. pipes was closed such that the neighborhood was only being served by the 8-in; the valve off the 24-in. was mostly closed. A crew put a valve key on the valve and said, "It looks like it's open all the way." And began walking away. What could a modeler do?

A good modeler might want to ensure that the crew is on the right valve, so have the operators run the valve up and down and count the turns. They made two turns and the valve seemed closed, but this was not the right number of turns for a 12 in. pipe. They were told to try to run it up again and when they got to two turns, they forced the valve, it went past the sticking point, and it opened as the 12-in. valve should open. Apparently, this 12-in. valve was open only 1 in. After the valve was fully opened, the model matched the field flow test.

Lesson learned: Valves may not be in their correct position.

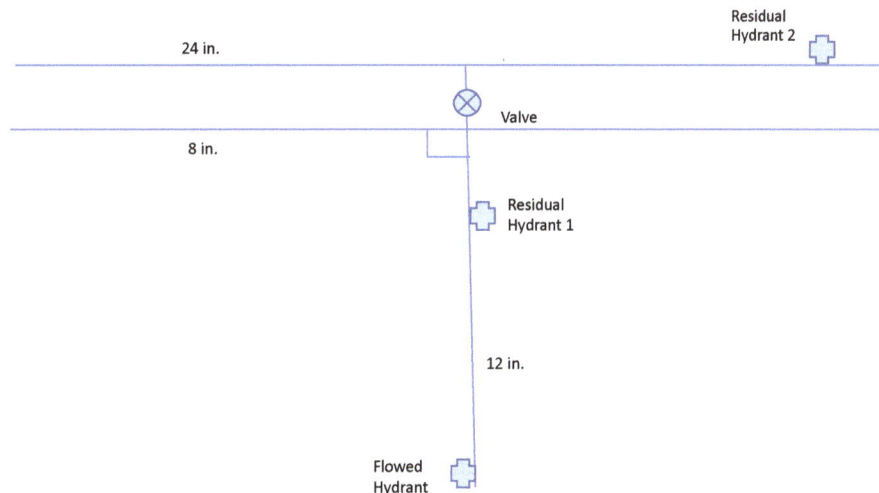

Source: Courtesy of Tom Walski

Figure 4-3   Schematic of hydrant flow test

Models have many different types of control valves. The choice of a control valve in a model may not always be the same type of valve that exists in the field, as the best model representation of a system may not necessarily be an exact match to field operations. Often, when the model has too many valves in a system, the model may not converge, and the valves may need to be simplified, using a single valve to represent multiple valves. Also, during calibration, the model may find that a control valve setting is incorrect and will need to be modified appropriately. When a model setting of a control valve disagrees with the value in the field, the question often arises as to which is correct. Should the model be adjusted to match the field value, or should the valve setting be changed to match what operators thought it should be when building the model? At times, a control valve may slowly stray from its setpoint depending on the type of valve. Operations personnel should be consulted here to provide insight on the control valve so that a modeler can appropriately represent it and its setting. Additionally, minor losses may need to be added to control valves to model situations where the minor losses govern over the valve settings.

Case Study: Pump Station Flows Don't Match Due to Missing PRV

A hydraulic model was being calibrated for EPS. One particular pressure zone was thought to be served by a single booster station and an elevated tank. The tank levels could not be matched well, and it was discovered that the field data showed that the tank continued to fill although the pump station had turned off. During the calibration process, it was discovered that there was a 2-inch PRV which also fed this pressure zone from the higher zone. This PRV station was not modeled, and therefore the tank levels were not calibrating because the flows from the PRV were not accounted for in the model.

Lesson learned: Missing facilities, even small ones, can alter system hydraulics and make model calibration a challenge.

Pressure reducing and sustaining valves. Pressure reducing valves are used to maintain downstream pressure. Setpoints for these valves may need to be modified during calibration based on field data.

Flow control valves. Flow control valves (FCVs) are used to control a set flow rate in a model. These are often used by modelers to set boundary conditions for flow; it is important to remember that the upstream head of any valve must be higher than the downstream head for the valve to flow. Modelers have sometimes used negative demand nodes to represent flows; this method is not recommended. FCVs will still allow for flow variation based on HGL in the distribution system; negative demands do not allow for this type of variability in operation. The use of negative demands or FCV should only be used as a modeling shortcut unless a specific condition is best represented by them.

Throttle control valves. TCVs are used to represent head loss through a valve within a distribution system, where a relationship between minor loss (k-value) and percent open is specified by the modeler. Automatically controlled valves are often challenging to model, in which case it will be important to understand and represent the logic in the programmable logic controller (PLC) that operates the valve.

Float valves. Float valves are used to control the filling of tanks based on the tank level. A float valve requires a flow versus pressure drop curve.

Losses through treatment systems or hydroturbines. Occasionally, treatment systems or hydroturbines need to be included within distribution system models. Depending on the type of system and how it is controlled, they could be modeled using any of the different control valve types available as part of the selected modeling software. The control valve type needs to be selected carefully to match the hydraulics of the treatment system, to the degree necessary, and the treatment system's impact on the water distribution system.

Considerations for Calibration During Steady-State High-Flow Conditions

Frequently, a model may appear to be calibrated for normal flow conditions when it contains errors that only become apparent when velocities and head losses are large. Because system design is based on worst-case conditions such as fires, peak days, and shutdowns, it is essential that the model be calibrated over a range of conditions from a low-demand day to reasonable high-flow conditions. For systems where fire flows are much higher than normal demands, head loss on normal days may be so low that errors in head loss between the source and the measured point within the distribution system may be larger than the head loss itself. Therefore, adjustments made to calibrate the model to match field data (e.g., changes to roughness coefficients) may not be advisable under those low-flow conditions.

Pipe roughness. Roughness coefficients can be modified during calibration. There are industry-referenced values for roughness coefficients based on material, pipe size, age, and condition that are used in the absence of information. Indeed, for many systems, sufficient information to set roughness coefficients for particular pipes is rarely available. The roughness coefficients can be calculated by a fire hydrant flow test for small - and medium-sized pipes, targeting specific pipe materials and/or areas of the system; transmission main monitoring during high flows can be used for very large pipes to estimate appropriate roughness coefficients.

In setting up a model, the modeler typically has two choices in what head loss equations to use: Hazen-Williams and Darcy-Weisbach. Hazen-Williams is an empirical formula based on field observations and has historically been used more frequently due to its ease of use. Hazen-Williams friction factors (C-factors) have been published in many sources and vary by diameter, material, age, and condition of pipe. These factors can be adjusted during calibration since the actual roughness coefficients are commonly unknown. It is important to recognize, however, that the Hazen-Williams equation is defined to be accurate under a certain regime of Reynolds numbers, pipe sizes (4-in. to 60-in.), and for clean water at typical temperatures (4–20°C). Darcy-Weisbach is also in common use, though less frequent than Hazen-Williams. It is more theoretically correct than Hazen-Williams, but the equation sometimes takes model software more time to solve compared to Hazen-Williams.

The importance of pipe roughness values depends on the pipe velocity and type of pipe. Tuberculation in old unlined cast iron pipes tends to show the greatest range of roughness values, although scaling can be an issue in other types of pipes as well. Pipe roughness tests (as described in Chapter 3) can provide a range of values which can be used to correlate the year a pipe was installed with the roughness values. By correlating year installed with roughness, a table can be developed to represent that relationship, and these values used in as part of the initial attempt at model calibration. Starting with reasonable values in such a table may lead to minimal adjustments needed to achieve calibration due to roughness.

Additional adjustments made to pipe roughness should be reasonable during the iterations of calibration. For example, if a very low C-factor (e.g., 20) is needed to match field conditions, it is usually a sign that there is a closed valve in the area. While if a very high C-factor is required (e.g., 200), it may be an indication that there is a missing pipe or connection or an incorrect diameter (e.g., a 6-in. pipe in the model in a run of 24-in. pipes).

Closed isolation valves. It is important to know the status of isolation valves and other maintenance work, to the extent possible, during the time of model calibration. Closed or partially closed isolation valves (where they should be open) and open isolation valves (where they should be closed) often cause differences between model results and field measurements. Additionally, and unless these periods can be avoided, any maintenance work ongoing during the time of the calibration must be reflected properly, such as flushing programs, valve exercise programs, and other field work, which can have impacts on model results.

When conducting hydrant flow tests in an area where closed valves are suspected, it may be helpful to install multiple data loggers in the area. If, during a hydrant flow test, the pressure does not drop at one logger but drops significantly at the next logger downstream, it may be an indication of a closed valve between the two.

A similar test can be conducted along a large transmission main, even when the pipe is so large that a hydrant flow test does not create a significant head loss. During periods of high flow, one could measure the hydraulic grade at number of locations along the transmission main then plot the hydraulic grade profile as a function of distance from the source. An abrupt decrease along the profile could indicate a closed or partly closed valve while a generally steady slope could indicate issues with pipe roughness coefficients.

Case Study: What PRV?

In a remote section of a large water distribution system, the hydraulic model could match fire flow results well, but during low-flow periods, the measured pressures were well below the observed pressure. What was causing the pressure drop during low flows?

The modelers could not develop an explanation until they had a meeting with operations staff who worked in that area. One said something to the effect of, "We used to have really high pressures out there until we put that pressure reducing valve (PRV) along highway X." The modeler's response was, "What PRV along highway X?" It didn't show up on system maps. When the PRV was added to the model with the correct settings, the model and observed pressures matched, while during high flow the PRV opened wide and those results still matched.

Lesson learned: Check system information with operations personnel to ensure the model reflects real conditions.

Spatial demand allocation. Demands need to be allocated to nodes as part of the model development process. If an EPS model is to be developed and calibrated, then diurnal patterns need to be developed, often based on system data, if available.

Demands in the model should represent the amount of demand in the field during the time of the field test. Total demand should account for production, wholesale supply or demand, and non-revenue water, not simply billing or account metered data.

Demand allocation is important and can make a significant difference to calibration. The allocation can be done at a very high level but will depend on the use of the model. Sometimes modelers can spend a significant amount of time allocating demands at a level of detail that doesn't noticeably improve calibration. Identification of large users and the specific pipes they draw demand from can be important as part of calibration.

As mentioned, it is important to define actual demands at the time of the calibration. For example, during a hydrant test, if the flushing is performed in the morning versus the middle of the day, demands can vary significantly and impact head loss and operations. Demands have daily, weekly, and seasonal variability, and sufficient data need to be collected and allocated to the model to match the calibration conditions. Large users should be placed individually in the model. Non-revenue water in the form of leakage must be accounted for in the model. If a large demand needs to be added to an area to achieve good agreement with field data, it can be a sign of a large leak or a closed valve.

The advent of automated meter reading (AMR) makes it easier to represent demands accurately in a system. However, it remains a challenge to spatially allocate non-revenue water (e.g., leakage), especially in systems where this is a significant portion of the total demand. Systems that are divided into district metered areas (DMAs) make it much easier to identify issues and spatially allocate such demands.

Case Study: Wholesale meter(s)

A large city sold water via wholesale meter to an adjacent utility district. The district was located near the boundary between a low and high-pressure zone in the large system. The model showed that the utility district was fed from the high zone. With the demand from the district, however, the model was predicting more pressure drop (lower HGL) than observed in the field. What could explain that?

Checking the billing records of the utility district showed that the wholesale meters from the high-pressure zone reported zero at one date and after that date, some new wholesale meters from the low-pressure zone appeared. Apparently, the operators had decided to feed the utility district from the low-pressure zone and did not tell the modelers.

Lesson learned: Modelers and operators need to communicate extensively.

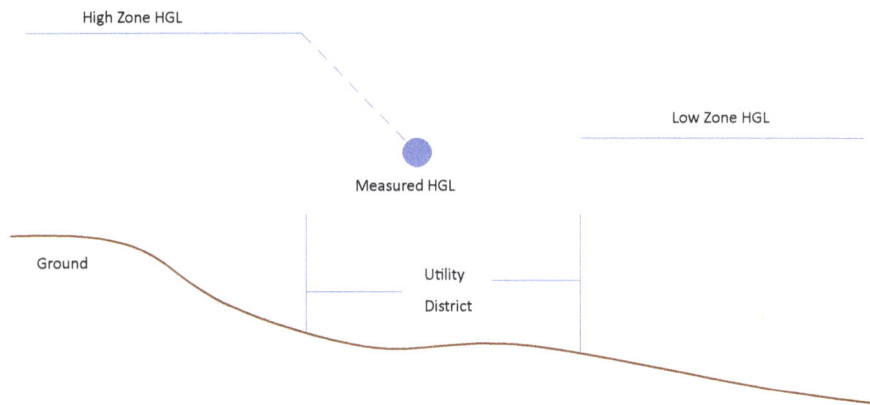

Source: Courtesy of Tom Walski

Figure 4-4 Profile view of hydraulic grade lines (HGL)

Adjustments to demands may need to be made during calibration due to their impact on system hydraulics. Common errors in the allocation of demand include the following:

- Incorrect spatial allocation. The location of largest customers should be confirmed and allocated as accurately as possible within the model so that their demands are drawn from the appropriate pipe(s) and pressure zones as necessary for users with multiple points of supply.
- Incorrect temporal factors – hourly basis. Hourly demands need to match the period of calibration. If a model is calibrated under steady-state conditions (e.g., to represent a hydrant flow test), then the system demand should represent that specific snapshot of time.
- Incorrect temporal factors – seasonal basis. The model demands should reflect the demands at the appropriate time of year for which the calibration occurs.
- Use of events and holidays in demand calculations. Often special days, such as Thanksgiving or the Super Bowl, have unusual demand patterns. Unless there is specific interest in investigating one of these days or data are most complete under those periods, these events are best avoided for calibration.

■ Errors in customer information database. Inevitably, there are errors in the billing databases used for demand estimation and allocation. If there are some unusual demands (especially high ones), there is the possibility of meter read errors in the billing database that have not been corrected. Some systems may have gaps in their billing data that are spatially non-uniform, meaning some areas may not have demand reported when demand in fact exists. For the purposes of model calibration, these areas should be more closely evaluated to determine whether there are potential issues with the database used.

■ Neglecting to include non-revenue water. After the database is reviewed for accuracy and completeness, non-revenue water (e.g., leakage) must be accounted for in the model. This is commonly approached through distributing the difference between supply and demand, after accounting for tank levels, uniformly throughout the network unless advised otherwise through discussions with system operations staff.

Case Study: Pressure Dependent Demands

Most of our hydraulic models are demand driven; that is to say, that demands are placed in the model, and it is assumed that the water system is always able to meet those demands. There are systems where the pipes do not always remain pressurized throughout the whole system (often in developing countries) or where system pressures fluctuate greatly (sometimes in irrigation systems) where the water system cannot be assumed to always be demand driven. In this case, models must be built using pressure dependent demands, where the demand modulates by pressure. Some of the modeling software packages offer this feature, which will allow for modeling intermittent demands.

Lesson learned: If a system is only partially pressurized, different modeling techniques must be used.

Connectivity errors. A model network is only as accurate as the maps and GIS used to create it. A map of the system may look adequate for purposes of general reference, but may contain errors, depending on the source, that could impact the accuracy of the hydraulic model developed to represent the system.

Incorrect connectivity at intersections or pipe crossings. Incorrect connectivity at intersections is often a common error in hydraulic models, which may need to be adjusted during calibration. At some intersections, there are often pipes that cross over other pipes without connecting, while at intersections there are some that do connect. In some CAD or GIS data sets, at a tee intersection, the through pipe is continuous, and therefore the pipe at the tee does not connect. Ultimately, if the original CAD or GIS data do not correctly represent these intersections, then it is likely the model is also incorrect.

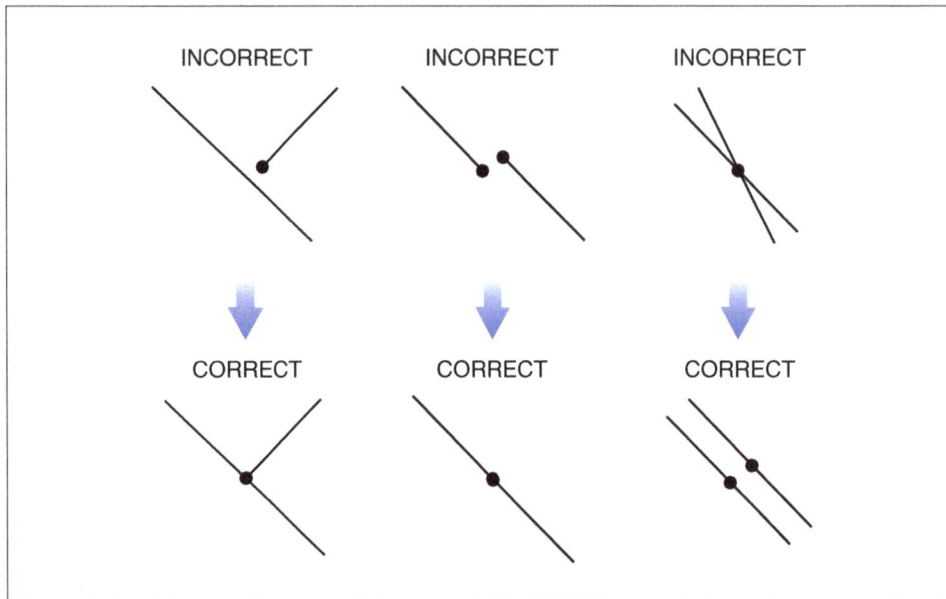

Source: Carollo Engineers, Inc.

Figure 4-5 Illustration of incorrect and correct placement of nodes

When models are built, model nodes are often created based on a certain threshold distance from the end of the pipe. If the ends of two sections of pipe are too far apart in the original data compared to the threshold, even if they are supposed to connect, then the model will create two separate nodes. If too large a distance is used as a threshold for connecting two pipes, two unconnected pipes may be represented as connected in the model.

Some modeling software contain tools to help identify potential problems with connectivity such as two nodes that are very close to one another and should have been connected, a node that is placed very near a pipe and should split the pipe into two, and a pipe that has been entered twice such that there are two identical pipes connecting a pair of nodes. Network tracing tools and color coding of pipes can highlight such problems. If pipes are color coded by flow, a very low flow in a transmission main can indicate connectivity errors, as can flow not taking its expected path such as a small pipe with a very high velocity.

Errors in original CAD/GIS data. In addition to linework errors in the original CAD or GIS data, data fields may also be incorrect. Incorrect diameters can cause significant model calibration problems, for example, an entry of 16-inch diameter might mistakenly be input as a 6-inch diameter, which would cause a significant head loss during model calibration.

Pipe sizes

Diameters. In most cases, models use nominal diameter of pipelines in the model. In reality, nominal diameters may or may not match internal diameters, especially in the case of HDPE pipe. Some modelers choose to use internal diameters rather than nominal diameters if the data are available.

Lengths. Some software packages automatically calculate pipe length, scaling it to the length of the line in GIS or other coordinate-based model software. Errors occur when the modeler thinks that pipe lengths are scaled and auto-calculated when they are not. Errors can also be caused when the model is not scaled correctly to the coordinate system. These length errors will need to be adjusted during calibration, if they exist.

Excessive Skeletonization. Historically, many models have been skeletonized, meaning that only major pipelines and facilities were represented in the model due to limitations in computing power, model software, or intended use of the model. Skeletonization is the process of eliminating small or short pipes and combining parallel or serial pipes into a single pipe segment. If a model has too few pipelines, it may not adequately reflect the water system, and pipelines may need to be added during calibration. Whenever possible models should include all watermains in a system; computing power has limited the need for skeletonization, and most models run very quickly. However, as a general rule, models should include all pipe loops to the extent practicable.

Some skeletonized models may have been "over-skeletonized" such that they misrepresent the connectivity and system hydraulics. If the model is only intended to be used for the transmission system, such misrepresentations may not have a significant impact, while for some purposes such as water quality or flushing planning, the model will need to be adjusted by including additional system detail to be useful. Additionally, C-factors in skeletonized areas may need some adjustment to represent hydraulic grade and pressure within those areas.

Case Study: Skeletonizing Flow Tests

The skeletonized model of a large urban system match flow tests fairly well. One exception was a neighborhood with a large number of small pipes which had been removed during skeletonization. During normal demands, the model agreed with the static pressures.

However, the model predicted more pressure drop than was observed during a hydrant flow test. As some of the small pipes near the flowed hydrant were added to the model, the model and the flow test began to converge. The more pipes that were added, the better the agreement.

Lesson learned: Models of high flows where significant head loss occurs won't be very accurate in the vicinity of the test when important pipes, especially looped pipes, are removed from the model due to skeletonization.

Considerations for Calibration During Extended Period Simulation Conditions

Before moving on to calibrating an EPS model, it is best to first calibrate the model under steady state conditions, and resolve potential issues associated with pipe roughness or closed valves, connectivity, pump curves, and elevation representation. Otherwise, under longer duration simulations, it may become more difficult to identify the likely cause of differences between modeled and measured values.

Usually, EPS calibration is based on data from the SCADA system. However, SCADA data are only collected at few discrete locations in the distribution system, if at all. Relying solely on SCADA data for calibration means that the model may be well calibrated from the source(s) to specific locations where recorded measurements are made (e.g., tanks), but may be inaccurate for other locations in between. Data loggers can provide data (e.g., pressure) for key locations not covered by the SCADA system, and less expensive sensor technology coupled with software and data access and analysis may increase the broader use of sensors in the future.

In general, for EPS calibration, tank water levels (and pump flows to a lesser extent) are the key data to compare, and graphs of model vs. actual tank levels are often the starting point for such comparisons. As a rule of thumb, if the graphs of tank level curves are in the same direction but the slope differs between the model and the SCADA data, the controls are good, but demands may need adjustment. If the slopes are in the wrong direction, then the controls are the source of the problem.

In EPS models, errors in demand, pipe roughness, or control setpoints can lead to major time shifts as to when pump status, or valve status where applicable, changes. These are almost impossible to avoid unless time-based controls for these pumps or valves are used. It is much more important that the pump flows are correct than matching the exact pump on/off times. As mentioned in Chapter 2, and elaborated further here, time-based controls (i.e., when controls are known) should be used first to calibrate a model under EPS conditions before applying specific rules (i.e., unknown controls) to minimize unknowns.

Pum and valve control. Depending on the complexity of the water system and the level of operator modification of automatic controls, the SCADA data may not appear to exhibit a predictable cycle or pattern of pump or valve starts/stops. The modeler should be able to pick out the portion of the data that represent preset controls and attempt to make the model mimic those time periods. Excessive matching to complex, and non-predictable, time series data can make the model especially difficult to calibrate. If the modeler cannot determine a pattern of operation from the field data, operations personnel should be interviewed to determine how they make decisions regarding the turning of pumps on and off and changing valve settings. For systems that are manually operated, it may be necessary to initially use time-based controls to check on inputs such as demand patterns. Once those operations are better understood, conversations with the ultimate users of the model can determine how such controls should be modeled.

When pumps and control valves do not represent flow accurately, it is advised to resolve these differences during steady-state calibration in advance of performing an EPS calibration. If a pump does not represent measured flow rate, the pump curve may be incorrect. If specific pumps were not active during the steady-state calibration, but were during the EPS condition, then a more detailed evaluation of those pumps may be warranted. Another possibility could be with upstream or downstream demands, connectivity near the inlet or outlet, or pressure zone delineation depending on the location and extent of differences observed.

Incorrect on/off or pump speed settings. Pumps and control valves require correct status settings. These are on/off settings for a steady-state model or an initial setting for an EPS model. Pumps need to be set either on or off (although another control might change that status); pumps with variable speed drives might have a pump speed setting. Control valves need to be set between actively controlling, fully open, or closed. If these are not set to match field conditions during calibration, the model will not calibrate correctly. Also, there is often confusion between control valve settings where the valve is actively controlling based on a setpoint, and fully open, where the valve acts like a section of pipe with a minor loss.

Case Study: Pump Station Controls

A hydraulic modeler was performing an EPS calibration of a system. As part of this system, there was a pump station which pumped from a transmission pipeline to a higher pressure zone with an elevated tank. Initially, the modeler set up the pump station to be controlled by levels in the elevated tank, with the pumps turning on at low tank levels, and turning off at high tank levels. During calibration, however, the modeler found that the pump station controls were more complex than merely turning on and off by tank levels. In reality, the pumps were also controlled by discharge pressure (turning on at low discharge pressure and off at high discharge pressure) and suction pressure (one setpoint which blocked other pumps from turning on and a lower setpoint which turned all pumps off). It was not until all of these controls were added to the pump station in the model, did the model reflect field conditions of pumps turning on and off.

Lesson Learned: It is important to understand field controls and operations when modeling facilities. Simplifications made for models may create erroneous results.

Complex and multiple controls. Pumps may have more complex or multiple controls (e.g., based on time delay, discharge or tank levels, suction block or stop, other pumps running). One of the most important decisions that need to be made during EPS calibration is these control settings. To the extent possible, the modeler should use controls that reflect field conditions but also are general enough that the model can be run under a range of conditions. In an ideal case, controls are based on level, pressure, and/or flow. The modeler must understand these physical controls, then decide how to model the controls and whether the controls should be simplified. Pumps and control valves may be controlled by multiple elements or field values; the modeler must decide which of these to include in the controls and may choose to simplify some of them.

There are some systems, however, that are operated manually, which makes it difficult to use logical controls during calibration. Time-based controls can be used for calibration to reduce the uncertainty around automated or rule-based operations. The risk of using time-based controls is that the model will be less suited to run under a variety of conditions, and therefore, if possible, should be minimized when the model is used for system evaluation. Use of time controls during calibration rather than rule-based controls means that if rule-based controls are added for system evaluation, a check will need to occur to confirm that the rule-based controls will perform equally well when they are used.

Emergency controls. Sometimes, emergency controls are activated by events in the field such as g fire hydrant flow tests. Sometimes pumps and/or control valves are turned on because of the high demands associated with the flow test. It is important to note which facilities turn on so that the same can be reflected in the model.

Case Study: Pump and Fill Valve Controls

Common to many water systems, especially those in flat areas, are ground level storage reservoirs where the tanks are filled from the system and pumped back to the same system. Complex controls are necessary for this type of system to match system operations. Of utmost importance are controls which do not allow the pumps to operate simultaneously with the fill valves. In this case study, an agency had a situation where there were five pumps from the ground storage tank, as well as one fill valve from the system. The fill valve operated during the nighttime and mid-day hours, while the pumps operated during both morning and evening peaks. The pumps all had the same setpoint to turn on at 63 psi and all turned off at 78 psi, while the fill valve was set to maintain a system pressure of 72 psi. To model this correctly, when the first pump turned on, the fill valve was turned off. If the first pump was running, and the pressure remained below the low pressure setpoint of 63 psi, then a second pump would turn on, which the subsequent five pumps following the same pattern. If the pumps were running and the pressure was above 78 psi, then the pumps would turn off, one at a time in the model. Finally, when the last pump was turned off, then the fill valve would be turned on at its setpoint. This is an example where complex controls were necessary in the model to appropriately match field conditions, requiring a number of conditional statements for those controls (i.e., IF and AND statements).

Lesson learned: Valve controls can be complex and must be well understood to model them correctly.

Case Study: Manually Operated Valves

One of the most challenging parts of a water system to model is a percent open control valve, where the operators modulate the percent open status of a valve to reduce pressure or control flow. These valves are often used to fill ground level storage tanks but can also be used from one pressure zone to another. In the case that field data is available for flow, suction and discharge pressure, and percent open, these valves can be modeled fairly well using a throttle control valve or some similar valve type. But when full data is not available, these valves often require modeling using some other type of valve – often flow or pressure control, which might not be the accurate type of valve in the field but gives the model the most realistic results. This is an example of a time where the modeler needs to select between matching the element type exactly versus matching the available field results more accurately.

Lesson learned: Field information is often not available for control valves and modelers must apply their engineering judgment and discussions with operators to represent the system appropriately to yield results that reflect actual conditions.

Demand patterns. The pattern of demands in an EPS model reflects demand over the course of a day. Ideally, the patterns used should best reflect actual field conditions during the day or time of calibration and can be calculated from SCADA or other field data. If the resulting patterns look erratic for the day of calibration, then it's advised to review the data over a few days with similar demand and operating conditions to determine whether some modifications to the demand patterns ultimately applied to the model are warranted. Reference or typical diurnal patterns should not be used during calibration wherever possible, as these might not reflect actual field conditions, and therefore, may result in a poor match between model results and field data. If diurnal patterns are for an entire system or multiple pressure zones, which is often the case, then the diurnal curve may not reflect operations in a particular portion of the water system where water use is dissimilar from other portions of the system. These can be adjusted, while ensuring that demand patterns are normalized, to better match tank turnover or pump operations.

Tank dimensions. Incorrect tank dimensions will give erroneous results during an EPS model calibration.

Tracer study. Tracer study data, such as the tracking of fluoride through a water system, can be useful for hydraulic calibration as it can give an indication of travel times, flow rates, and supply from one source over another system wide or to specific pressure zones. This is discussed in further detail in Chapter 6.

Validation. Once a model is calibrated, it is customary to validate the model by running it under a different data set (e.g., different demand or operating condition) than the one used for calibration. Like calibration, boundary conditions need to be set prior to model validation. This will often produce results with a greater difference than observed during the calibration conditions, not because the model is inaccurate but because no two days are identical in a water system. There may be different demands, shutdowns, or operator decisions that can change the results from the model. The model should not be expected to match these observations perfectly, but the modeler should be able to explain the differences. Some adjustments to the model may be warranted to improve the match to these multiple conditions.

AUTOMATED CALIBRATION

Some software contains tools that can make "automatic" adjustments to parameters to potentially achieve better calibration results. Such tools can enable models to match pressure and flows at specific points in the system. These tools can be useful when one has many pressure data points in a single pressure zone and when seeking guidance on how to approach manual adjustments; for example, an automated routine that identifies where pipe roughness must be excessively reduced to improve calibration may help identify the location of closed valves. Regardless, the user must be sure that the tool is adjusting the right parameter, and to within a reasonable range. Otherwise, the calibration can be achieved by compensating errors where an error in demand can hide an error in pipe roughness. In using such a tool, the head loss measurements must be significantly greater than the error in head loss measurements or else the results amount to calibrating measurement errors. The results of automated calibration should be verified using a separate set of field data.

CALIBRATION DATA REPORTING

Although calibration result might not specifically be part of a planning or design report, they can be helpful for reference, There are several options to report calibration results and compare modeling results with field results, which may include mean, standard deviation, or even percent difference.

After the data are collected and the total demand in the model is equivalent to total production on the target day, the results of the model are compared with field data. Depending on preferences, tools, and the software available, several methods of comparison are available.

1) **Spreadsheet based**: Electronic spreadsheets can be used to compare model output against field data for multiple iterations of the calibration process. Data from each source can be copied into the spreadsheet and plotted

for each data stream. Model data can be stored in a separate spreadsheet or file and just be referenced in. In this way when changes are made to the model, the references can be simply refreshed without having to copy and paste repeatedly.

2) **Commercial calibration extension**: Some commercial modeling packages include modules to assist the modeler in the calibration process. Some of these software packages include automated features that attempt to facilitate the process of adjusting model parameters to arrive at a good level of calibration. As mentioned previously, these tools require a minimum level of understanding of how the software functions and how to avoid potentially unsuitable calibration results.

3) **Custom programming**: Some hydraulic modeling software is based on open-source programming codes (such as EPANET) that make programming iterative simulations and adjustments possible. Modelers skilled in programming can create tools customized to each model that can aid in the calibration process.

CHAPTER 5

Real-Time Modeling and Continuous Calibration

Jim Cooper, Rajan Ray, Luke Butler, Lindle Willnow

Water utilities have and are continuing to make investments toward digital transformations, including SCADA (Supervisory Control and Data Acquisition), Internet of Things (IoT), Geographic Information Systems (GIS), and hydraulic models. In the past, these technology platforms have been siloed within specific utility departments resulting in only a select few utility professionals able to access and benefit from the data. With SCADA, there can be gigabytes of data representing years of pressure, flow, water quality, and tank data that are never accessed. The technology and data fidelity within these systems have continually improved, providing more confidence with the data coming in from the field and recent enhancements allow for better data access and integration options across these systems. This enables utilities to leverage these information technology (IT) investments to support the development of real-time models to accurately predict and forecast system hydraulics on a continual basis. Furthermore, these real-time models can be leveraged to validate and troubleshoot the IT resources providing data to the models.

Most hydraulic models are calibrated based on using short-term samples of field data for both steady-state and extended period calibration. These calibrated models have proven to be useful tools for a wide array of applications from system planning to pressure zone management. That stated, calibration being based on a specific date/time range of data may not accurately represent the system conditions under the wide-varying range of operating conditions that may occur. A real-time model, which can be continuously compared with data provided by SCADA and/or other field sensors, provides more frequent comparison points and thus provides more realistic predictions based on the varying operating conditions.

A model becomes real-time when connected to the utility's SCADA system and/or other real-time sensor data in a continuous fashion such that boundary conditions (e.g., tank water levels) and operational statuses (e.g., pump speeds or on–off status, valve settings) can be applied to the model with little or no manual intervention. Real-time models can be programmed to run "automated," which refers to simulations that are run on a regular basis without human intervention in order to update the model based on real-time boundary conditions and perhaps forecasted system demands.

The synthesis of real-time data with simulation models provides the ability for utility operators to perform applications beyond traditional planning functions. Operators can routinely engage in situational operational training to look at incident response (such as pipe, valve, and pump failures), monitor pump energy on a frequent basis, troubleshoot water quality issues, optimize system operations, evaluate leakage, and more. Benefits being that the utility now has tools that both engineering and operations can use to streamline operations and improve service to their customers. In terms of calibration, the overall benefit of implementing a real-time model is that model calibration is streamlined and more frequent than traditional non–real-time models. The outcome is a model that is continually being used for predictive modeling and is continually validated, in turn, providing a high level of confidence for operators and engineers to make reliable, informed decisions.

CONTINUOUS CALIBRATION/CALIBRATION WITH REAL-TIME MODELS

When it comes to frequency of hydraulic model calibration, there are different requirements based on the model's scope and scale of use. "Continuous calibration," defined further below, provides a level of calibration that suits most modeling applications whether it be for fire flow simulation, water quality analysis, energy modeling, master planning, etc. Continuous calibration is the process of periodically comparing a hydraulic model to real-world data and adjusting the model accordingly. The types of adjustments that can be made vary and include control modifications, demand multiplier or pattern changes, status changes, or even roughness adjustments. Automation of these adjustments will depend on the utility and their history with the model and their SCADA system. The frequency is set by the modeler and can occur on a nearly continuous basis (e.g., every 15 minutes, 1 hour, 1 day, etc.) or on a more periodic timeframe such as monthly, bi-monthly, or yearly. As shown in Figure 5-1, the model is populated by reading SCADA data at the desired frequency, updating the network model boundary conditions and operational statuses, pausing execution and generating the corresponding network analysis results, and then waiting for the new SCADA measurements to reload the network model and rerun the network simulation. The model results can then be compared to the SCADA data to validate the level of calibration.

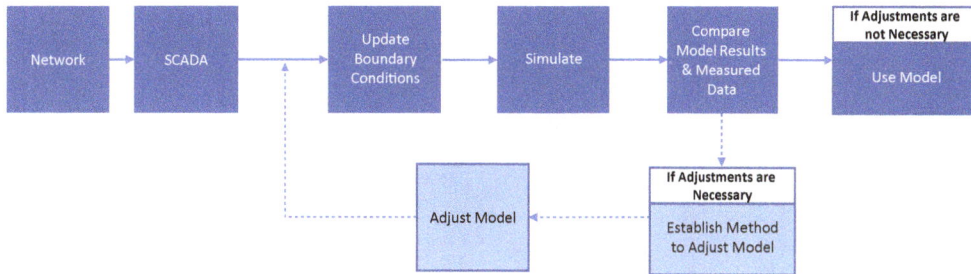

Figure 5-1 Continuous calibration process schematic

The Trends in Water Distribution System Modeling survey (Committee Report, *Journal AWWA*, Oct. 2014) reported that one of the most technically challenging aspects of hydraulic models (42 percent of all respondents) is calibration. A quarter of the respondents indicated that their hydraulic model has not been calibrated in more than five years. Utilities typically hire a consultant with experience in building and calibrating hydraulic models every few years to update and recalibrate their model. These costs can be considerable when needing to collect new data from Utility Operations, perform field calibration tests, validate facility conditions, prepare calibration reports, model deliverables, etc. Many times during the calibration process, the consultant may find valves that were left closed in the field months, perhaps years ago and/or pump settings that are outdated in the model, and similar issues. These are all valuable items to find and adjust, but what if they could be found sooner? There are hidden costs associated with the timeframe of finding these discrepancies. Not to say there are no costs in setting up a continuously calibrated model but perhaps that option should be evaluated by the utility and/or their consultant and compared to the traditional calibration over the long term.

Aside from potential cost savings by reducing the frequency and scope of traditional model calibration, the accuracy and quality of continuously calibrated hydraulic models can be significantly improved when compared to the traditional calibration process provided that sufficient quantity and quality of live data streams are available. With a continuous calibrated model, the user has the assurance of performing analysis on a well-calibrated model using current data. Consider the confidence level on performing modeling what-if simulations based on today's real-world data vs. a model that was calibrated 5 years ago based on data sampled during several days (usually in a peak demand condition) with a discrete operating condition. These continuously calibrated models have a direct effect on quality control and quality assurance in planning and operations as well as building confidence in the model. Improved confidence in model results provides a pathway for more value across the utility particularly in Operations and Management. Operations staff frequently adjust operations of the distribution system based on daily variables including field work, weather, water quality, energy costs, etc. so a predictive model becomes an important tool when deciding what and when to adjust, and the hydraulic impact. This affects the safe, efficient delivery of water which is a top concern of utility management. By routinely comparing

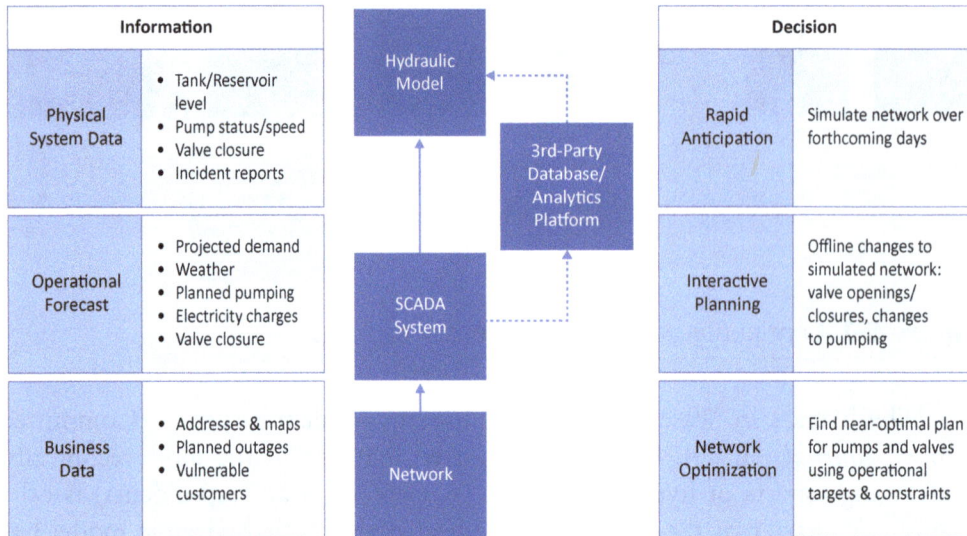

Figure 5-2 Systems information and decision opportunities with continuous calibration

the model results to real-world data, one can find network issues sooner than traditional methods of learning about these problems (such as customer complaints, field investigations, and similar). For example, if a tank is draining much faster than being reported in the model, that may point to a valve that has been left open or potentially a water main break within that zone. Again, this is benefit for operations that have reliable model results in locations between where they have data being retrieved from SCADA monitoring points. These model results provide a more comprehensive understanding of the impact of their operational decision-making. Additional opportunities are identified in Figure 5-2.

To develop a continuously calibrated model, one must connect the hydraulic model to the utility's SCADA database (typically using a backup server of the SCADA system called a "historian"). There are several modeling software packages that offer the ability to connect to these databases directly. Another route is to go through an intermediate dataset which can be as simple as a comma-separated values (CSV) file or a data analytics platform that imports SCADA data (among other different datasets) as it is collected and performs various data processing and analysis on the data. The concept of connecting is the same where the modeling software connects to the dataset and updates the model's boundary conditions (e.g., tank water levels) and operational statuses (e.g., pump speeds or on–off status, control valve settings) with little or no manual intervention.

As described in previous chapters, well-calibrated models result from how well the model was originally built and updated. Network accuracy, demand allocation, diurnal curves, pump curves and settings, operational controls, etc. affect the hydraulic results ultimately impacting the model calibration. For example, when representing pumps in a model, one typically provides the specific pump curve supplied by the pump manufacturer; however, when that pump deteriorates over time– this change is seldom updated within the hydraulic model.

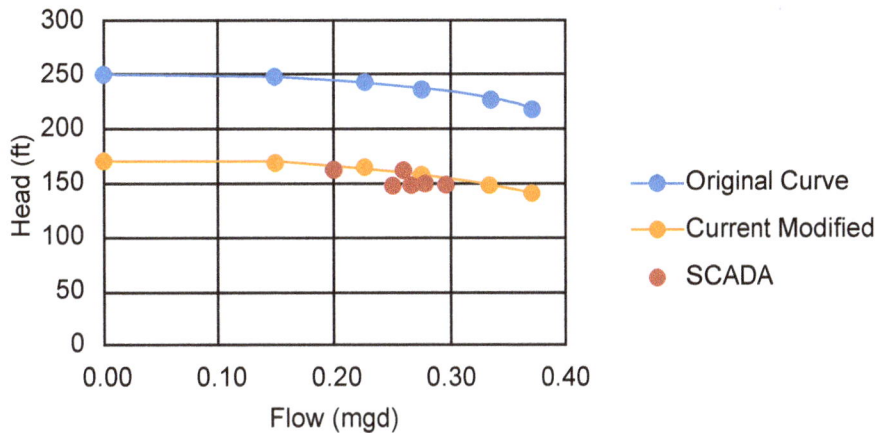

Figure 5-3 Comparison of operational performance and modeled pump characteristics

Such examples, as shown in Figure 5-3, have a direct impact on the hydraulic results and would reflect a discrepancy when compared with the real-world data points. One of the by-products of implementing a continuous calibrated model is the ability to analyze pump operating points to identify which pumps are not performing per their original performance curves. Advanced applications of continuous calibration may result in periodic adjustment to the pump curves accordingly. The other method of validating pump curves is to perform pump tests in the field which is much more resource-intensive vs. evaluating the pump performance via the associated pump SCADA output.

Reflecting pump and control valve settings (when they turn on/off or are regulated) is a common task when developing and calibrating hydraulic models. By integrating the SCADA facility points to the model for continuous calibration, is much more streamlined as the modeler can more readily develop and then compare the model operation set points to the real-world SCADA graphs. For example, by viewing a pump station outflow graph in relation to the tank level chart, one can easily identify when the pump station turns on/off based on the inflection points of the tank level chart. This also provides the ability for modelers to quickly identify/adjust any of the model's operational controls when the values of the model results deviate from the real world. Another benefit is the ability for a utility to develop automated controls from their current manual operational controls.

Understanding demands and related patterns and how these impact the trends of critical infrastructure including tanks and control valves is important in system insight and model calibration. By having a continuously calibrated model, one can validate these usage patterns on a more frequent basis. Also, having a continuously calibrated model implies that your model is connected to SCADA so it provides the data points required for developing a diurnal curve, a common step of performing an hourly mass balance using production, imported water supplies, pump flows, and change in storage. By having access to this data, one can more readily model variations of these patterns as they relate to seasons, weather patterns, weekday vs. weekend usage schedules, conservation efforts, etc. Another development in real-time modeling is linking AMR/AMI (Automated Meter Reading/Infrastructure) data to the models so that the demands can be updated on a more frequent basis following additional data processing and setup. AMR/AMI data can also be used to fine-tune demands for larger users or patterns for different user categories.

Continuous calibrated models shift the traditional focus of modeling from planning to operations and provide a reliable tool for utility operators to help operate their systems more proactively. This is advantageous for the utility as it promotes further use of the model as confidence in the results increases. In operations, this can mean utilizing the model for day-to-day analysis. SCADA is inherently a tool to understand what is happening at specific monitored locations at the current time. The hydraulic model provides insight by interpolating hydraulic properties between those points. For example, a decision may be made to turn off a pump feeding a tank. With a continuous calibrated hydraulic model, the operator can view the status of pressures, flows, and demand in the surrounding pipes within that zone to aid in timing of their decision. This is also useful for field activities (such as hydrant flushing, valve closures, and related) where operators need to know status of system pressures and flows to reliably perform their tasks without a significant impact on their customers. The main aspect that significantly assists operators beyond traditional SCADA systems is a model's ability to forecast and perform what-if simulations. This is helpful in many areas including identifying efficient operation schedules based on varying demand conditions, pump optimization per energy tariff schedules, emergency response scenarios, operator training, and related areas. Also, by having the model results compared against SCADA and field monitor data on a frequent basis, operators can identify deviations that can point to leakage incidents, failed assets, monitors that need maintenance, valves that were left open/closed in the field, pumps acting off their pump curve, and so on. With any hydraulic model, confidence leads to more frequent utilization, use and advantages to utility operations team, and advanced applications of the tool. Continuous calibration is a process that utilities can implement to provide consistent and reliable model results to help grow that level of confidence.

DATA INTEGRATION AND DATA QUALITY

Automating the calibration of a hydraulic model requires integration between the utility's model and its SCADA system.

In general, the model uses SCADA data for the following purposes:

- Setting initial status of pumps (on/off/speed), valves (open/closed/position), and reservoirs (level)
- Establishing time-based control rules for pumps and valves or for confirming that the automated controls in the model match the actual system controls

▪ Comparing with model results (pressures, flows, levels)
▪ Calculation of total consumption within the model and modification of the demand and demand patterns

While a water utilities SCADA system contains this data, much of the data requires pre-processing before it can be used within the model or used in comparison with modeling results. This pre-processing can include data aggregation, validation, and quality checks. Ideally, the data stream feeding the model only contains data used by the model. Filtering of extraneous readings not related to the model may be necessary. The SCADA data interval should also match the time steps used in the model.

In a traditional model calibration, this data is sourced once, typically manually, during the initial calibration. If the modeler identifies any data anomalies, fixes to these issues typically occur outside of the telemetry source on the extracted data.

A real-time model needs access to high-quality data continuously with no manual intervention. With automated access to this telemetry data, verification of values calculated within the model and those in the SCADA system is possible during each scheduled model run.

This section describes the pre-processing used for different types of data and discusses special considerations taken for some of the data and the steps that are required to ensure the storage of only quality data within the utility SCADA system.

A real-time model that is continuously running needs access to current SCADA data in near real-time to set boundary conditions, verify model results, and predict future values, as shown in Figure 5-4.

Getting access to data in near real-time can be challenging as water utilities will typically firewall their SCADA systems from their corporate networks as a security measure to avoid unauthorized access. Coordination with the IT network

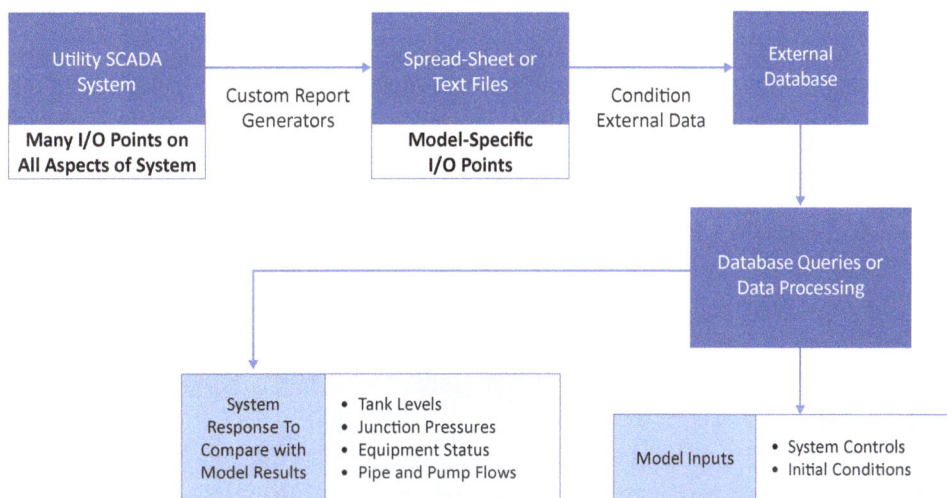

Figure 5-4 Data flow and processing in a continuously calibrated model

security team will likely be needed to ensure continuous access to the data required for the model. Some utilities use data historians to bridge the firewall and act as a read-only view into the operation of the network.

However, historians, depending on their administration, can have a limited data set compared to what is available directly from the SCADA system. Custom reports can be developed to ensure that the data required by the model, and only that data, are made available to the model.

Data Feed Schedules and How Often Updates Are Made

The primary source of network geometry and the related attributes required to build a model is the utility's GIS system. While some GIS systems contain layers that provide a linkage to the SCADA system, the depth and quality of this information can vary. While some may have a one-to-one link between features and SCADA points, in many cases, there exists a one-to-many relationship, and because of this, GIS may display limited information. Furthermore, some systems only link the existence of SCADA at a facility, or some may have no link at all and require manual searching and retrieval.

The IDs used within a model often come from the GIS primary key, and while some more organized utilities may share the same keys between their GIS and SCADA systems, typically, they are different and require lookup to match the two data sources.

The development of a mapping table between the model IDs and SCADA IDs is essential for a real-time model. While initially, the source of some of this data may come from GIS, further work would be required to fill in gaps. While some SCADA data may be directly usable in the model, some preprocessing may have to occur before it is usable. These can be simple tasks such as matching units and time series intervals between SCADA and the model. Or can become more complex such as validating data to find outliers or bridge gaps, and cleaning data by inferring or combining multiple SCADA tags. The real-time model will be running on a continuous stream of data from SCADA and ensuring that this data is correct is key.

Incorrect data can cause the model to either not converge or provide erroneous results because of conflicting "knowns" or it can cause the model to force calibration intervention into issues that do not exist. An initial first clean up and audit of data is required. As the model moves into a continuous calibration, the quality of SCADA data should stay at a high standard. The ongoing running and validating of the data will provide continuous feedback on the quality of the SCADA data.

The SCADA tags needed as a minimum to set the initial boundary conditions of the model are tank levels, pump on/off status, and control valve positions. To verify the model, measured flows, pressures, and levels should be used as a minimum.

It is generally expected that model elements with attached live data should have corresponding information for each time step within an extended period simulation. However, gaps can occur, and backfilling these by utilizing some of the methods described on the next page should be considered.

- Data with a jagged profile, specifically those that influence demand and cause instability within the model can be smoothed using a moving average
- Small outliers can be identified and corrected with reasonable minimum and maximum values or highlighted if they deviate too far from the previous value. These can also be averaged out
- Where small gaps occur, historical averages on similar days can be utilized
- Where more significant gaps exist, offsetting the time to get a similar weekday profile can be completed. If flows are significantly different, then results may not be as expected and therefore identifying a similar day is important.
- More sophisticated methods such as machine learning and AI can be used to backfill data gaps

It is important to note that the SCADA data may not always be reliable. One benefit of using a continuously calibrated model is that the model results can help identify when an instrument is providing readings out of line with readings from other system sensors. By following trends in the readings, instrument errors such as a slow drift in meter accuracy can be identified and addressed. There may be instances when an instrument is returned to service after an outage event, like a power failure, that readings could be adversely affected.

While a traditional model may only try to replicate an average day or peak day scenario, the real operations of a pumping station can be more complex and subject to change depending on local conditions. By continuously validating the status and operation of pumps and valves within the model against the near real-time data of SCADA provides confidence that the automated controls in the model match the actual system controls. Automated controls can be validated at their simplest by ensuring that pumps are cycling on and off as predicted. Further discussion on pump controls and evaluating SCADA data for calibration can be found in Chapter 3.

Pressures within the network can be continuously monitored. While most pressure monitoring would occur at facilities, additional monitoring at hydrants and other critical locations within a pressure zone can allow further analysis of results. In some cases, pressure data can be taken at customer meters vastly broadening the ability to continuously calibrate.

By comparing the extended period simulation results against pressure data in the middle of a zone, deviations in pressure can identify the following issues:

- Point restrictions within the network, such as unintentionally closed or partially closed valves
- Increase in pipe wall roughness due to age
- Localized high usage, either legitimate or unauthorized
- Significant leakage or watermain breaks.

Depending on the data sources, the capabilities for updating frequency are wide-ranging with examples of utilities updating the model every 5-minutes based on their SCADA data updates to once every 6 months based on seasonal demand updates. The frequency of updates depends on the intended use of the model. Using real-time models for operational applications versus longer-term planning studies will require a higher frequency of data updates. For example, the frequency of model updates when evaluating a pump scheduling for energy optimization could be run each morning, so operations staff can plan how to operate based on energy tariff charges and storage requirements. Other utilities may want to update models based on a significant hydraulic change to their system (i.e., a pump station upgrade, new development, substantial change in usage, etc.) where the real-time model could be updated in a less-frequent manner.

REAL-TIME MODELING CALIBRATION AND INTELLIGENT WATER

Computer modeling of water distribution systems has occurred for decades. Over time, there have been significant enhancements in software, hardware, communication, and information management processes. These enhancements have allowed engineers and planners to ask new questions, including different and more complex questions such as obtaining a live pulse on the system during a large demand event or other special operating condition. While modeling of water distribution systems relies on a calibrated model to best-represent current system conditions, many models today continue to be based on average or typical system conditions with specific scenarios developed to represent a typical minimum or maximum day event. Water systems are now realizing the potential benefits of real-time modeling for improved operations, among other applications. Models that utilize live data streams require ongoing adjustment to replicate the systems in near real time. Ongoing advancements in information management and modeling will result in more and more models operating as real-time models.

Intelligent water is a common term used to describe the process of water systems embracing digital technologies and ecosystems in front-line operations and in utility management with the purpose of improving financial stability, customer experience, and operations and maintenance key performance indicators. This is fueled by a cultural shift toward innovation. Intelligent water is also important to meet evolving customer and workforce expectations. Many customers value technology and may expect their water system to have real-time awareness at the individual customer service levels and to be able to access their own data in real-time, especially as the cost of water increases. The population is migrating toward a technology-rich environment where digital technology is woven throughout their daily life. They are comfortable with and expect a technological evolution with constantly improving solutions. Thus, utilities with Intelligent water technologies can better compete with other industries in attracting talent and further empower their workforce to make well-informed decisions. Real-time models, digital twins and live connection to system data all align with the process of systems advancement toward an intelligent water future.

Case Study: Las Vegas Valley Water District (LVVWD)

The real-time modeling process at LVVWD allows for rapid identification of model and field issues. The LVVWD reservoir level comparison tool is frequently used for this purpose and can generate SCADA-versus-forecasted results. Monitoring SCADA in comparison with model predictions allows anomalies in the system to be rapidly identified, investigated, and corrected. In some instances, energy savings are realized not only by optimizing the pumping schedule but also by correcting field conditions that are costing money. For example, a reservoir water-level tracking report at the district's 2745 North Reservoir showed that the reservoir was filling faster than the model had predicted (Figure 5-5a). In this case, the problem occurred in an area of heavy development activity. The investigation found three zone isolation valves were open, conveying water to lower-pressure zones and filling the 2745 North Reservoir. When the valves were closed, the reservoir tracking returned to normal (Figure 5-5b). This is a case in which the LVVWD operational modeling program quickly drew attention to a field issue that might otherwise have gone unnoticed for a long time at significant cost in wasted energy.

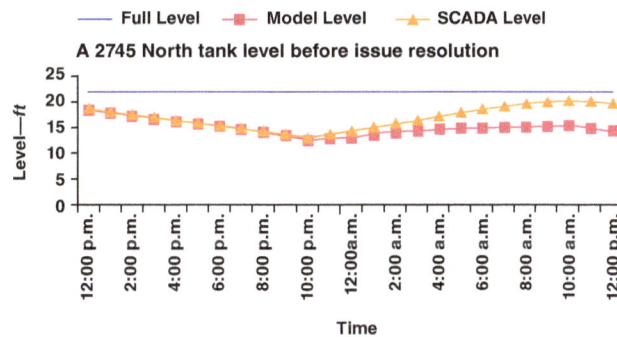

Figure 5-5a A report showing that the reservoir was filling faster than the model predicted

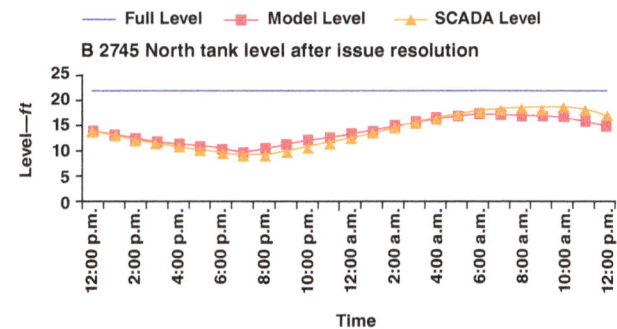

SCADA—supervisory control and acquistion

Source: P. F. Boulos, L. B. Jacobsen, J. E. Heath, and S. Kamojjala, "Real-time modeling of water distribution systems: A case study," Journal AWWA, Vol. 106, Issue 9. E391-E401.

Figure 5-5b Reservoir tracking returned to normal after valve closure

Lesson Learned: Resolving field issues is streamlined with real-time modeling

While this chapter intends to summarize real-time model calibration, there is significant advancement currently underway. The following examples demonstrate the direction of water distribution system modeling and potential for future applications.

■ **Shift in User Experience** – Most models today are tied to a local computer and software on a local machine. An opportunity to shift from desktop computer-based to cloud-based or mobile-based modeling applications can be combined with advancements in augmented reality to simulate and visualize anticipated conditions for field crews in real time.

■ **Enablement of Digital Twins** – As levels of digital twins begin to emerge, a common initial level is simply a virtual representation of physical assets. While this aligns with today's hydraulic model, as data stream integrations become commonplace, the model-will be a fundamental component in the development of a system-wide higher-level digital twin.

■ **Processing Enhancements** – Advanced modeling analyses can be quite time intensive today, particularly with the potential complexities of a Multi-Species Extension (MSX) simulation. The EPANET algorithm is limited to single core processing because the subsequent iteration relies on the results of the previous one. Improvements to the algorithm may open water distribution modeling to multi-core processing for standard simulations.

■ **Integration of Artificial Intelligence (AI)** – AI is only beginning to impact the water sector. A future exists where water distribution modeling will benefit from AI in ways such as automated calibration and live data quality checks for field data utilized for model calibration. This will enable live comparison of model results to field data for proactive notification of system operational anomalies.

While the fundamental hydraulics of a water distribution system have been well understood for many years, the future of modeling continues to evolve and will continue to evolve over time. Model use cases continue to change as software and hardware enable increasingly advanced capabilities. Any future use cases must begin with a calibrated model that is representative of the conditions the user attempts to simulate.

REFERENCES

Boulos, P.F. & Jacobsen, Laura & Heath, J. & Kamojjala, Sri. (2014). Real-Time Modeling of Water Distribution Systems: A Case Study. Journal - American Water Works Association. 106. E391-E401. 10.5942/jawwa.2014.106.0076.

Huaqiang, C. (2016). Research on Real-time Hydraulic Modeling and Applications in Urban Water Supply Network DMAs.

Sunela, M.I.; Puust, R. (2015). Procedia Engineering. 119: 744-752.

Uber, J.G.; Boccelli, D.; Woo, H.; Su, Y. (2013). Real-Time Network Hydraulic Modeling: Data Transformation, Model Calibration, and Simulation Accuracy. Somerset, Ky.: National Institue of Hometown Security.

CHAPTER 6

Water Quality Calibration

Walter Grayman, Ferdous Mahmood, Meg Roberts

Water quality modeling involves the simulation of the fate and transport of substances within the distribution system. It can provide information that is useful in both the design and operation of water distribution systems to ensure acceptable water quality being delivered to customers. The water quality modeling umbrella includes simulation of constituents in the water, source tracing, and water age. A well-calibrated hydraulic model is a prerequisite to the calibration and application of a water quality model.

HISTORICAL PERSPECTIVE

The development of water quality modeling dates back to the work at US Environmental Protection Agency (USEPA) in the early 1980s on water quality in distribution systems (Clark, 2015). The first such models were steady state models (Males and Clark, 1985), followed by the development of dynamic water quality models that were used in conjunction with extended period simulation (EPS) hydraulic models (Grayman et al., 1988; Liou and Kroon, 1987). In 1991, the AWWA Research Foundation and USEPA held a technology transfer workshop in which experts from around the world discussed the state of the art and the future of water quality modeling including a session on calibration (USEPA, 1991).

Major areas of research, development, and application related to modeling of water quality in distribution systems modeling have included mixing and water quality in tanks (Grayman et al., 2000); fate and transport of disinfectants and disinfectant by-products (Vasconcelos et al., 1997); contamination modeling associated with waterborne outbreaks and source water contamination (Grayman et al., 2004); and water security issues including source identification and siting of monitors (Janke et al., 2014). Recent developments in this area include the Multi-Species eXtension (MSX) to EPANET that facilitates simulation of more complex substances (Shang et al., 2011), and real-time modeling. Calibration is an important aspect of applications of models in all of these areas.

Relationship Between Hydraulic Modeling and Water Quality Modeling

Water quality modeling is done in conjunction with hydraulic modeling of the distribution system. The hydraulic model provides flow solutions used as input to the water quality model which subsequently uses the hydraulic data to calculate fate and transport of the substances. While early water quality models were separate programs from the hydraulic modeling software, today the two components are integrated into a single package.

A basic assumption of most water quality models is that "dissolved substances travel down the length of a pipe with the same average velocity as the carrier fluid while at the same time reacting (either growing or decaying) at a given rate" (AWWA M32, 2018). Most water quality models also assume that at junctions receiving inflow from two or more pipes, the mixing of fluid is complete and instantaneous.

Calibration of Water Quality Models

Because of the tight pairing between hydraulic and water quality models and the reliance of water quality modeling on the hydraulic model results, a well-calibrated eps hydraulic model is an essential element for performing water quality simulations. Generally, the level of calibration required for a hydraulic model is stricter when used for water quality modeling than for many other uses. Water quality modeling requires full knowledge of the actual paths, flows, and travel times that water follows as it moves through the network. Use of a hydraulic model that has been improperly or incompletely calibrated, will invariably lead to poor water quality simulation results. Panguluri et al. (2005) contain a detailed discussion of calibrating water quality models.

Beyond the hydraulic calibration process, there are two purely water quality characteristics that can be adjusted during the water quality calibration process: reaction types and coefficients and mixing characteristics in tanks.

Types of Water Quality Modeling and Calibration Needs

There are three general types of modeling that are included under the water quality modeling umbrella: water age, source tracing, and constituent modeling. All three types of modeling are available in most software packages.

Water age. Water age is a measure of the flow-weighted travel time from all sources to any point within the distribution system. It is implicit within water age calculations that: (1) water age follows zero-order kinetics with a rate constant equal to 1, whereby each minute that passes after water enters the distribution system, the water becomes a minute older until consumed by demand; and (2) water age can be linearly combined when two pipes come together. For example, if two pipes flow into a node with equal flows, then the resulting water age in the downstream pipe is the average of the water age in the two upstream pipes flowing into the node.

There are no reaction coefficients that can be adjusted when modeling water age and, consequently, the accuracy of a water age model is highly dependent upon the accuracy of the underlying hydraulic model. Water age can be strongly influenced by the time that the water spends in a storage tank as well as the mixing characteristics of the tank. Therefore, adjustment of the mixing regime model and mixing coefficients for tanks need to be examined when calibrating a model used for calculating water age. This also applies to source tracing and constituent modeling.

Tracer studies in a distribution system, using a conservative tracer such as fluoride or calcium chloride, are useful in calibrating a water age model by comparing the travel time of the conservative tracer to the model predictions of water age. This topic is discussed further later in this chapter.

Water age is frequently used as a surrogate for water quality because water age is directly related to disinfectant decay, disinfection by-product formation, and many other biological and chemical reactions.

Source tracing. Source tracing is another form of water quality modeling. In source tracing, the model calculates the percent of water over time that reaches any node in the network with its origin from one specified node. The source node acts like a constant source of non-reacting constituent that enters with a concentration of 100. Like water age, calibrating a hydraulic model for purposes of source tracing requires an accurate hydraulic calibration. The mixing model and mixing coefficients for tanks should also be calibrated during the calibration process. This type of modeling is most useful in water systems that have multiple sources with differing water quality signatures (Hatcher et al., 2004).

A tracer study is also useful when calibrating a model used for source tracing. The conservative tracer is generally injected at a single source and the resulting concentrations monitored over time throughout the distribution system.

Constituent modeling. A constituent in a water distribution system can be classified as either conservative (non-reactive) or non-conservative (reactive). The concentration of a conservative substance does not change except when additional quantities of the constituent are introduced into the distribution system or when flow in two pipes with differing concentrations comes together. As a result, there are no reaction coefficients associated with conservative substances. Beyond the hydraulic calibration, the only calibration parameters for modeling conservative substances are those that describe the mixing characteristics within storage tanks. Tank mixing is discussed later in this chapter.

For non-conservative substances, in addition to calibrating the parameters discussed above for conservative substances, the form and coefficient parameters associated with the reactions are needed. Modeling constituents involves predicting growth or decay via chemical and/or biological reactions. Conservative constituents assume no growth or decay and as a result are a function only of mass transport in water. Non-conservative constituents are typically modeled by either empirical or mechanistic models.

Empirical models are based on direct observation, measurement, and extensive data records. Mechanistic models are based on an understanding of the behavior of a constituent and the constituents with which it interacts. Most water quality non-conservative constituents have empirical representations and/or a combination of mechanistic and empirical models.

Free chlorine, for example, is often modeled with a first-order decay reaction:

$$C = C_o e^{(-kt)}$$

where:

C_o = initial concentration at head of pipe (mg/L)
t = time spent in the pipe (h)
k = first-order chlorine decay constant (L/h)

This reaction is often further refined by variable decay rates in the bulk water and the water along pipe walls. Wall decay may be needed to accurately represent the decay and is very site specific where:

$k = k_b + k_w$
k_b = bulk decay constant
k_w = wall decay constant

Other non-conservative constituents of concern in water distribution systems include chloramine, disinfection by-products (DBPs), corrosion by-products, and arsenic as well as other potential contaminants. These constituents have unique and more complicated reactions than first-order decay and depend on other non-conservative constituents. Multi-Species eXtension (MSX) modeling can help the modeler to develop predictions for constituents that involve multiple interdependent parameters as discussed in the next section.

The challenge with modeling non-conservative constituents is determining the appropriate model formulation and coefficients. A statistically significant number of measurements must be taken to calculate the formula coefficients. In the chlorine example above, the bulk decay coefficient is usually determined by jar (bottle) tests in the laboratory (presented in more detail later in this chapter). The wall decay coefficient is more challenging and is most often determined by using the hydraulic model with a trial-and-error approach using a given set of distribution system measurements of chlorine (presented in more detail below).

Calibrating non-conservative constituent models requires measurements of the non-conservative constituents in the distribution system and comparison of these measurements to model predictions.

Some non-conservative constituents, such as DBPs, are more accurately predicted using a power function in lieu of an assumption of first-order formation. In such cases, mathematical methods (and software) can be used to determine equation coefficients.

Multi-species eXtension (MSX). Water quality modeling software provide a set of predefined methods for modeling non-conservative substances such as a first-order decay model that is frequently used to model chlorine residual. These pre-defined methods have been found to be adequate in many applications. However, it does not allow the modeler to use other fate relationships or to use fate relationships that involve multiple interdependent parameters. That constraint has been removed by recent developments in modeling software that allow users to define and model complex reaction schemes between multiple chemical and biological species in both the bulk flow and at the pipe wall (Shang et al., 2011).

The USEPA developed the Multi-Species eXtension to the EPANET engine, called EPANET-MSX (Shang et al., 2011) to provide greater flexibility in modeling a variety of water quality situations. For each application implemented under MSX, the user or developer specifies the mathematical equations that describe the reactions. Examples of processes that have been modeled using MSX include the auto-decomposition of chloramines to ammonia, the fate of DBPs such as HAA5, biological regrowth including nitrification dynamics, combined reaction rate constants in multisource systems, and mass transfer of limited oxidation pipe-wall adsorption reactions. The EPANET-MSX capability has also been incorporated into commercial packages.

Development and subsequent calibration of the water quality routines implemented under MSX are the responsibility of the developer/user. Software algorithms developed to model multiple species may employ a variety of mathematical relationships that are applicable to different types of constituents or combinations of constituents. The user must specify the mathematical expressions that govern the reaction dynamics of the system being studied, which affords users the flexibility to model a wide range of chemical reactions of interest to water utilities, consultants, and researchers. Development of new MSX routines adds a level of complexity in the modeling process and may entail a significant level of research. MSX has been used to model many water quality relationships such as arsenic (Burkhardt et al., 2017), chlorine (Grayman et al., 2011), and microbial contamination (Helbling and VanBriesen, 2010).

Comparison of types of water quality modeling. The type of water quality modeling that should be used depends on the question the modeler is asking. Many distribution system water quality challenges exist (AWWA, 2017). Compliance with the four major regulations (as of 2020) that impact the distribution system is of top concern to most utilities. These include the Revised Total Coliform Rule (RTCR), the Surface Water Treatment Rules (SWTRs), the Disinfectant/Disinfection Byproduct Rule (D/DBPR), and the Lead and Copper Rule (LCR).

The RTCR is intended to improve public health protection by setting maximum contaminant levels for E. coli to protect against potential fecal contamination. The SWTRs require utilities to check residual disinfectant concentration at entry points to the distribution system and at RTCR routine sites, and that 95 percent of these sites have detectable residuals (many states have more stringent requirements). The D/DBPR sets maximum allowable chlorine dose to the distribution system and requires utilities to measure total trihalomethanes (TTHM) and five haloacetic acids (HAA5), or chlorite and bromate for systems using chlorine dioxide or ozone for residual disinfectant, respectively, at specified distribution system locations which are unique to each system. The LCR was published to control lead and copper in drinking water. Lead and copper enter drinking water primarily through plumbing materials, but the chemistry of the water leaving the water treatment plant combined with the biological and chemical reactions in the distribution system have the potential to impact the water quality.

Other typical water quality concerns in the distribution system include biofilm, nitrification, taste and odor, discoloration, manganese, opportunistic pathogens, cross-connection control, and potential accidental or intentional contamination.

Water age is a surrogate for many of these water quality concerns, including disinfectant residual, DBPs, and nitrification. Source tracing is the best way to predict the area impacted by contaminations of various kinds introduced at the sources. In most cases, by assuming the constituent is conservative, the worst-case can be assessed. Modeling and predicting concentrations of other constituents can be very challenging due to the time and resources required to develop empirical models and their coefficients. Using water age and source tracing often can help utilities make informed decisions without investing excessive time and money into sophisticated water quality constituent models.

WATER QUALITY CALIBRATION PROCEDURES

Model Considerations for Water Quality Calibration

The following sections discuss the factors that need to be considered for a water quality model calibration.

Pipe network considerations. A hydraulic and water quality model may represent a distribution system's entire pipe network (i.e., "all-pipe" model) or only selected pipes (e.g., all pipes greater than a specified diameter or some other criteria). When only selected pipes are included, this is referred to as a "skeletonized model." Generally, for master planning purposes, a skeletonized version may be adequate to predict pressures and available fire flows and determine capital improvement needs. However, water age and water quality calculations can be very sensitive to smaller diameter dead-end pipes with little demand. The inclusion of smaller diameter pipes in a system can also change the water flow patterns which in turn can impact the water age/water quality characteristics in various parts of the system. Therefore, models used to determine maximum water age in a system or to calculate areas of low chlorine residual should include additional smaller diameter pipes (and especially dead-end pipes) all the way up to an all-pipes model. Very short lengths of pipes of the same diameter should be combined into single pipes so that the travel times in the pipes exceed the water quality time step in the hydraulic model.

Demand considerations. Typically, in distribution system modeling, demands are estimated based on monthly water consumption data from water meters, supplemented with temporal water use patterns derived from SCADA (Supervisory Control and Data Acquisition) data. While this method may be sufficient for master plan models, it may not be sufficiently accurate or granular to calibrate and model water age or water quality. Water quality models are dependent on knowing water flow patterns throughout the system, based on discrete diurnal patterns for water consumers that may vary both spatially and temporally throughout the water system. While modeling for hydraulic capacity is focused on peak demand periods, the critical cases for water quality modeling are often low-flow periods.

Techniques for generating more detailed consumption data include

- Use of Automated Meter Reading (AMR) or Advanced Metering Infrastructure (AMI) systems for developing diurnal patterns for large water users as well as other types of water users
- Assignment of DMA boundaries. By continuously measuring flow into and out of a DMA, a spatial record of water use in the DMA can be defined and used in the calibration and modeling process

Operational considerations. Operational controls are required in a hydraulic model to specify how selected components are operated. When used in water quality modeling, the controls can have a very significant impact on the ability to successfully model water quality. These controls are generally developed using a combination of SCADA data and discussions with system operators. Typically, repeating patterns of controls for specific seasons are incorporated into the model. The most common control feature is pump station operation based on storage tank water levels or pressure at specific locations within the distribution system. SCADA can also provide pump run times which may be incorporated into the hydraulic models. Other controls may involve set points (which can vary seasonally) for flow control valves and pressure reducer valves.

In many cases, operators run the distribution system on an ad-hoc basis. This method generally involves the operators watching the hydraulic characteristics of specific locations (such as storage tank water levels or system pressures) and adjusting controls (pumps or valves) to meet the desired hydraulic objectives. The inclusion of ad-hoc procedures into a water quality model is more challenging than the controls developed directly from SCADA. Given that operational controls of pumps, tanks, and valves have a significant impact on the flow patterns and ultimately water age/quality in a distribution system, the ad-hoc procedures need to be accounted for in a water quality model. Typically, the ad-hoc procedures are accounted for by discussing the system operations with the operators and combining the information gathered from SCADA to distill down operation controls into manageable repeating patterns.

During the water quality model calibration process, the general controls are sometimes replaced with actual records of pump off-on records and other controls to ensure an accurate representation of hydraulic behavior.

Valve status and percent closure. The open/close status of valves and their percent closure have significant impact on water quality model. In hydraulic modeling, incorrectly representing valve positions can lead to some errors in pressure calculations. However, in water quality modeling, errors in valve positions can lead to very significant errors in water quality calculations because it can cause large errors in defining the flow patterns.

A thorough discussion with system operators is needed to understand which valves are adjusted in the system and what system conditions drive those adjustments. Operations and maintenance (O&M) records may provide information about which valves were adjusted not just for seasonal demand variations but also for unusual system events such as line breaks and prior construction activities. Once the locations for the unusual system events have been identified, the presence of valves nearby and whether they were adjusted during these events need to be determined. The valves that were adjusted need to be reviewed to make sure that these valves are at their desired positions. One simple method that operators use to determine valve positions (open/closed/partially open) is counting how many turns were completed for the set position to fully open or close position. Knowledge of the valve size along with manufacturer's data sheets is used to determine percent open. These percent open values need to be incorporated into the water quality model. While such level of detail is typically not required for master planning model, it is recommended for water quality models. There are other methods to determine valve positions or set points such as flow and pressure measurements upstream and downstream of the valves, but these types of methods may require the closure of other valves nearby to isolate the effect of the valve in question.

The percent closure or set points of some valves can vary seasonally. Also, the flow through a valve varies with time of day. Different types of valves have different head loss characteristics as a function of flow. The valves that are intentionally partially open or the valves whose set points translate to partially open valves should have the appropriate valve characteristics in a water quality model. The valve characteristics are typically available from the manufacturers provided the size and type of valve are known. Having appropriate valve characteristics for partially open valves makes the water and source tracing more accurate in a water quality model.

During the model calibration process, discrepancies between measured values (pressure, flow, or constituent concentrations) and modeled values may indicate incorrectly set valve position representation and may suggest modifications needed in the model.

Calibrating tank mixing models. EPANET and most other hydraulic/water quality distribution system software packages provide the following four options for modeling how a water tank mixes:

1. Complete and instantaneous mixing
2. Plug flow: First in–first out (FIFO)
3. Plug flow: Last in–first out (LIFO)
4. 2-compartment mixing model

Each of these mixing models are idealized, theoretical representations of mixing models. In reality, no tank exactly follows any of these mixing models, so the calibration process involves selecting the mixing model that most closely represents the tank mixing behavior. With the exception of the fourth model (2-compartment model), there are no parameters associated with the mixing model. The default mixing model and the one that is most frequently used in modeling tanks is the complete mix model.

Following are some general guidelines that can be used in selecting the appropriate mixing model:

- A tank that experiences fill and draw cycles of several hours each with relatively robust inlet velocities (i.e., greater than a few feet per second) will generally mix well and can be represented by the complete mix model

- A tank with a separate inlet and outlet and baffles to control the flow paths between the inlet and outlet generally operate in a plug flow mode and the FIFO mixing model is most appropriate. Contact chambers at the end of a treatment plant are generally designed to operate in a FIFO mode

- Tall standpipes with a common large diameter inlet – outlet pipe at the bottom of the tank generally do not encourage good mixing. As a result, the last water entering the tank at the end of the fill cycle will be the first water leaving the tank at the start of the draw cycle. As a result, a LIFO model may be the best fit for this situation

- The two-compartment model has rarely been used in the past to model tanks and there is virtually no guidance in the literature for selecting the one parameter that must be specified in this type of model. The use of this model should be avoided unless there is specific field data, usually in the form of a tracer test, that indicates that this type of model is the best fit and could be used to parameterize the two-compartment model

Tracer tests (described in Section 1.2.2.2), either standalone tests for a tank, or tracer tests as part of one conducted for distribution system calibration, provide one mechanism for selecting/calibrating a tank mixing model. In such a test, the tracer concentration is measured, at a minimum, in the tank inlet and tank outlet (if there are separate inlets and outlets), or in the combined inlet/outlet. A more detailed study can also involve sampling at locations within the tank itself.

Another mechanism for selecting/calibrating an appropriate tank mixing model is the use of a computational fluid dynamics (CFD) model of the tank. A CFD model of a storage tank typically involves developing tank geometry and inlet/outlet pipe configurations from design/record drawings, developing a mesh for the geometry (computation cells), incorporating boundary conditions (inflow/outflow) from SCADA or operations data, and incorporating appropriate physics models that govern the hydraulics inside the tanks. A typical output from the CFD model showing mixing regimes is shown in Figure 6-3.

Sometimes the mixing characteristic of a tank is difficult to predict due to the combination of factors that affect the mixing, such as tank geometry, inlet/outlet pipe diameters and configuration, flow rates, water levels, and temperatures. In such cases, the mixing characteristics of a tank can be identified using the output from a CFD model as shown in Figure 6-3, which then helps to select the appropriate tank mixing sub-model or a combination of sub-models. Some examples

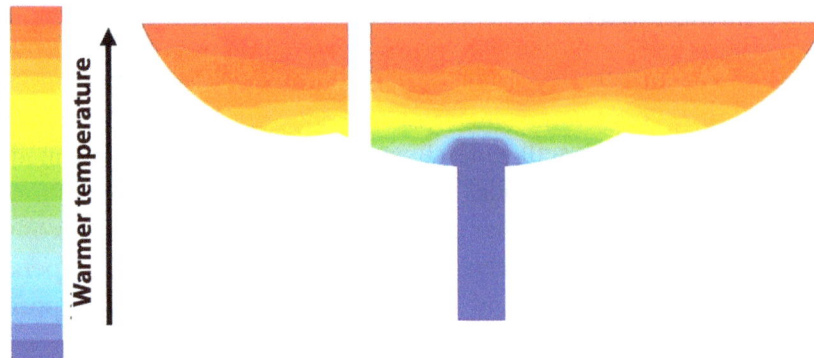

Figure 6-1 Output from a CFD model of elevated storage tank used to define mixing characteristics to select appropriate tank mixing model

may include tanks and or pipes in series or parallel to represent different mixing and water quality outputs.

Field and Laboratory Procedures

Data used in the calibration of water quality models is collected through a combination of field and laboratory studies. Collection and analysis of water quality-related data can be relatively expensive in comparison to hydraulic data. Therefore, care should be taken in planning and executing a field and laboratory program to stay within an allocated budget. The types of such studies and the procedures that are followed in performing these studies are discussed below.

Field test overview and procedures. The calibration of water quality models is largely dependent on the availability of water quality data collected in the field. That data may include measurements of water quality constituents that occur naturally in the raw water tempered by treatment works, by-products of treatment of the drinking water, or constituents that have been added to the water as part of a tracer test. In all cases, a well-designed program for collecting and analyzing samples is essential in order to create a data set that can be used with confidence in the calibration process. Some of the factors that should be considered in designing the sampling procedures include, but are not limited to:

- Sampling type: grab samples vs. automated continuous monitoring
- Monitoring/sampling locations
- Sampling frequency
- Sample analysis methods
- Proper use of monitoring instruments
- Logistical arrangements
- Safety issues

Additional information and guidance on water quality sampling can be found in AWWA's M32 Manual on *Computer Modeling of Water Distribution Systems* (AWWA 2018).

Tracer tests. A tracer study is a mechanism for determining the movement of water within a water distribution system through direct measurement of a naturally occurring or added substance within the water. In such a study, a conservative substance is typically injected into the water system and the resulting concentration of the substance is measured over time as it moves through the system. The results of a tracer study may be used as a mechanism for calibrating or validating a water distribution system model (Panguluri et al., 2005).

Tracer studies are used primarily to calibrate or validate the underlying hydraulic model itself, though the tracer test may also be used in support of a water quality study and to ensure that the hydraulic model is adequately calibrated to support water quality modeling. All water quality models of distribution systems depend upon hydraulic models to provide information on pipe flows, flow directions, and flow velocities. Inaccuracies in the flow/velocity values provided by the hydraulic model will lead to inaccuracies in water quality predictions.

Water quality models can be used to represent both conservative and non-conservative substances, but the use of conservative substances is most appropriate for calibration of hydraulic models. When modeling a conservative substance in a hydraulic/water quality model, parameter adjustments are made primarily on the hydraulic parameters rather than water quality parameters. The only major exception to this rule is calibration of the mixing model used when representing tanks (e.g., complete mix, plug flow, etc.). In other words, if the hydraulic parameters are correct and the initial conditions and loading conditions for the substance are accurately known, then the water quality model should provide a good estimate of the concentration of the substance throughout the network. The use of the water quality model and conservative tracer as a means of calibrating the hydraulic model is based upon this relationship.

Several conservative tracers have been used in water distribution system tracer studies. The primary ones are described below. When adding any chemical to a water distribution system or, in the case of turning off a fluoride feed, care should be taken to ensure meeting local, state, and federal regulations.

- Fluoride is a popular tracer for those utilities that routinely add fluoride as part of the treatment process. In this case, the fluoride feed can be shut off and a front of low fluoride water is traced as it moves through the system. A second tracer test (or a continuation of the initial test) can be performed when the fluoride injection is turned back on. Fluoride samples can be easily measured in the laboratory or ion specific electrodes (ISE) can be used in conjunction with data loggers as automatic monitors.
- Calcium chloride ($CaCl_2$) has been used in many tracer studies throughout the U.S. Generally, a food grade level of the substance is required. It can be monitored by measuring conductivity, or by measuring the calcium or chloride ion. Conductivity is the most suitable parameter when automated monitoring is used since inexpensive conductivity meters are readily available.

- Sodium chloride (NaCl) has many similar characteristics to calcium chloride in that it can be traced by monitoring for conductivity or the chloride or sodium ion. The allowable concentration for sodium chloride is limited by the secondary maximum contaminant level (MCL) for chloride and potential health impacts of elevated sodium levels in the water have been identified.
- Naturally occurring chemicals can also be used as tracers under some circumstances. This is most common when a system is fed by multiple sources with different chemical signatures such as differing hardness levels.

The calibration process of hydraulic and water quality models using tracer study results may be summarized as follows:

1. A conservative tracer is identified for a distribution system. Methods for injecting the tracer and monitoring it in the distribution system are determined.

2. A controlled field experiment is planned and performed in which either: (a) the conservative tracer is injected into the system for a prescribed period; (b) a conservative substance that is normally added such as fluoride is shut off for a prescribed period; or (c) a naturally occurring substance that differs between sources is traced. When injecting tracers, they may be injected as a single long pulse of several hours or as a series of shorter pulses.

3. During the field experiment, the concentration of the tracer is measured at selected locations in the distribution system using either automated continuous monitors or through manual samples that are analyzed in the field or in the laboratory. Additionally, other data that are required by a hydraulic model such as tank water levels, pump operations, flows, etc. are collected so that an accurate hydraulic model may be applied to the situation. In addition to the conservative tracer, other water quality concentrations may be measured such as chlorine residual as a mechanism for calibrating or validating the water quality rate coefficients.

4. The model is then run with alternative hydraulic parameter values to determine the hydraulic model parameters that result in the best representation of the field data. Perhaps, traditional pressure and flow measurements are used to perform a first-step calibration. The water quality model is then used to model the conservative tracer.

5. Good agreement between the predicted and observed tracer concentrations indicates a good calibration of

Fluoride for Station CORD, mg/L

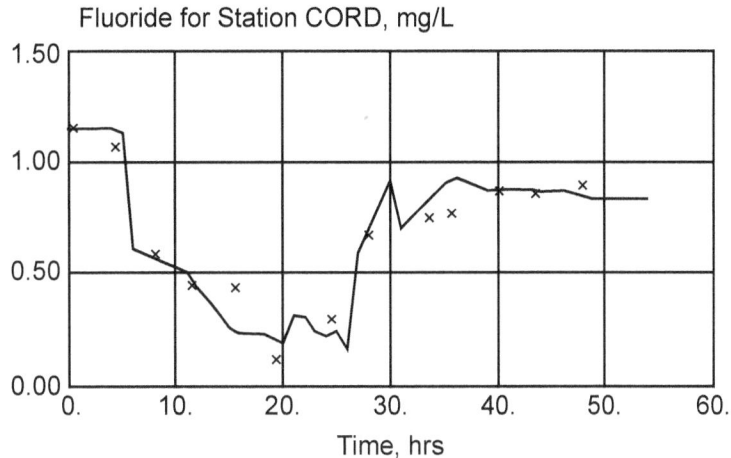

Source: J. Vasconcelos, P. Boulos, W. Grayman, et al. "Characterization and Modeling of Chlorine Decay in Distribution Systems." AWWA Research Foundation, Sept. 1996.

Figure 6-2 Example of plot of comparison of water quality data and model results

the hydraulic model for the conditions being modeled. The observed and modeled results can be compared either using a statistic such as root mean square error or through visual inspection of the graphs. Significant deviations between the observed and modeled concentrations indicate that further calibration of the hydraulic model or tank mixing models is required. Various statistical and directed search techniques may be used in conjunction with the conservative tracer data to aid the user in adjusting the hydraulic model parameters to better match the observed concentrations. The following diagram is an example of a comparison between the observed fluoride tracer concentration values (marked by X's) and the model results (shown as a continuous line). In this example, there is relatively good agreement between the observed and modeled results, suggesting a well-calibrated hydraulic model. If additional calibration is warranted, then an iterative process can be employed where hydraulic parameters may be refined and the modeled versus observed values displayed until an acceptable calibration is reached.

A case study describing a detailed tracer test used in calibrating a hydraulic/ water quality model is provided later in this chapter.

Bottle (Jar) Tests

Bottle (or jar) tests are a controlled laboratory-based mechanism for estimating the form and parameters governing the reaction of a water quality component over time due solely to the bulk characteristics of the reaction. Such tests are commonly performed to determine the reaction rates associated with chlorine decay and growth of disinfectant by-products over time. The exact details of a bottle test depend upon the specific component being studied. However, in most cases, several water samples are drawn at the finished water entry point to a water distribution system in separate bottles and stored at a fixed temperature away from light. Over a period of several days, at a pre-established frequency, a bottle is opened, and the concentration of the component being studied is measured. The results of the study are then plotted over time and the form and rate coefficients are determined.

An example of results for a chlorine decay bottle test is shown in the following graph. A first order decay curve with a decay rate of -0.1/day has been fit to the data.

An example of detailed procedures used in conducting a chlorine bottle test as delineated in the AWWA M32 manual is shown here.

Strategies for Calibrating Water Quality Models

Calibration of a water quality model involves several iterative steps. These steps are outlined below:

1. Ensure a well-calibrated hydraulic model for the conditions used in the water quality analysis. Typically, that may require a greater degree of calibration than is needed for most other uses of a hydraulic model or additional calibration for low-demand periods.

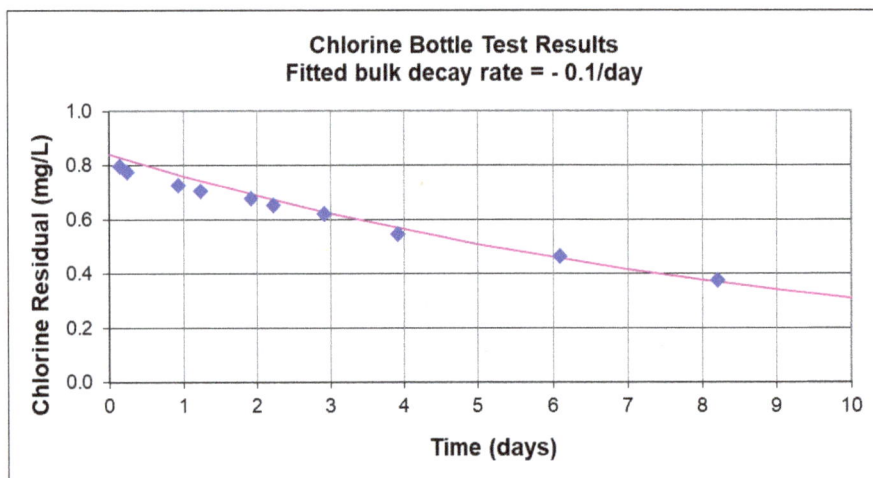

Source: Courtesy of Walter Grayman

Figure 6-3 Fitted chlorine decay curve

```
This procedure should be followed to generate data that can be used to estimate the
rate of chlorine decay in finished drinking water.

    1.  Split a single sample of finished water into 24 to 36 amber bottles of 250 ml
        size or larger. Fill the bottles so they are headspace free and cap them.

    2.  Place the bottles in a small bucket or wire cage and immerse them in a water
        bath in a sink with running water so that the initial water temperature of
        the sample can be maintained.

    3.  Starting from time zero, periodically remove three bottles at a time and
        analyze the contents for free chlorine. Discard bottle contents after taking
        measurements.

    4.  For each group of bottles analyzed, record the time (in hours from the start
        of the test), the free chlorine concentration of their contents, and the
        temperature of the water bath.

An ideal schedule for analyzing the bottles would provide at least 10 sets of
chlorine values that cover a range from 100% down to about 25% of the initial
chlorine concentration. A typical schedule might analyze bottles at 0, 0.5, 1, 2,
3, 6, 12, 18, 24, 36, 48, 60, and 72 hours. However some adjustments might have to
be made if the intermediate results indicate that chlorine is decaying either much
more rapidly or much more slowly than anticipated.

After the test data are generated, a bulk decay coefficient can be estimated by
plotting the natural logarithm of chlorine concentration versus time, and fitting a
straight line through the points. The use of three separate chlorine readings at
each time can be used to eliminate obvious outliers. The slope of the line is the
bulk decay coefficient in units of 1/time (e.g., if time is in hours then the
coefficient has units of 1/hours). When fitting the line it is best to force it to
pass through the initial measurement at time zero. Alternatively, one could use a
nonlinear curve fitting routine, available in many commercial spreadsheet and curve
plotting software packages, to estimate k_b directly from the equation
```

$$C = C_o \exp(-k_b t)$$

```
where C is the chlorine concentration measured at time t and C_o is the measured
concentration at time zero.
```

Source: AWWA M32, 2018.

Figure 6-4 Steps in conducting a bulk chlorine decay test

2. Collect field water quality data at selected sites in the distribution system. Samples should be taken during a period (or periods) representative of the time of year for which the model is to be calibrated. This may include continuous or grab samples over an extended period. Field data may also be required to determine the mixing characteristics in tanks and reservoirs.

3. Collect or assemble hydraulic-related information needed for the model such as pump operation, demands, tank water levels for the same period when the water quality data is being collected.

4. For reactive (non-conservative) substances, run laboratory bottle tests to determine reaction coefficients and appropriate reaction models.

5. Run the hydraulic/water quality model for the period of interest and compare the simulated constituent concentrations to the measurements made in the field.

Continue to adjust the model parameters to improve the agreement between measured and modeled results until you are satisfied with the match between modeled and observed values.

6. If the discrepancies between modeled and measured data are unacceptable, additional data collection may be necessary.

7. Model validation: Ideally, the model should be validated using an independent dataset that was not used in the calibration process. If the results of the validation step are not acceptable, then additional calibration may be required.

Calibrating a Chlorine Model

One of the most common uses of water quality models is the simulation of chlorine residual in a water distribution system. The purpose of such simulations is to help design or operate a water system so that a sufficient residual is maintained throughout the system at all times. There are many ways to calibrate a chlorine model but all of them depend upon the following general steps:

1. Ensure a well-calibrated and representative hydraulic model

2. Perform a bottle test to determine the chlorine bulk decay coefficient. If there are multiple water sources with significantly different chemical characteristics, then separate bottle tests should be run for each source.

3. Run the hydraulic/water quality model using the measured bulk decay rates and compare the simulated chlorine residual concentrations to chlorine measurements in the field. Adjust the model parameters to improve the agreement between measured and modeled results.

Generally, the biggest unknown in modeling chlorine is the wall demand exerted by the pipe on the chlorine concentration. Frequently in the initial simulations, the modeler assumes that the wall demand is zero. If the simulated concentrations are systemically higher than the field samples either for the entire system or in select regions, then the explanation could be that the unaccounted-for wall demand is a significant factor in the water distribution system. Direct measurement of the wall demand coefficient has not proven to be practical though some researchers have conducted experiments in this area. Use of an equation that relates the wall demand coefficient to pipe roughness has been suggested by some researchers and this capability is supported by most water distribution system software packages. However, there is little data available in the literature to parameterize such a relationship. The most commonly used approach for determining wall demand coefficients is the iterative adjustment of wall demand coefficients until an acceptable match is found between observed and modeled

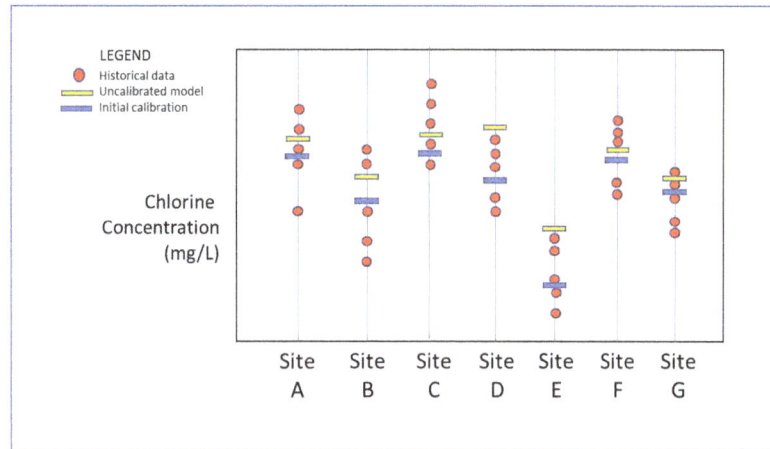

Source: Courtesy of Walter Grayman

Figure 6-5 Use of historical chlorine data in a first-cut calibration of a chlorine model

values. Wall demand is most significant in smaller, unlined cast iron pipes so these would be the first pipes that the modeler would examine in adjusting wall demands.

The degree of effort required for calibrating a chlorine model should depend upon the end use of the model. In an initial planning level study, use of historical chlorine data routinely collected as part of the TCR rule may serve as an initial step in calibrating the model. In the following example, the model is run under average demand conditions and compared to the range of historical data at seven sampling sites. The wall demand coefficients have been iteratively adjusted so that the model results reflect the general ranges of historical chlorine concentrations at each site.

For a more detailed study or even for refinements during a planning level study, field work is generally needed where chlorine data is collected throughout the distribution system over a period of a day or two. Chlorine samples can be taken manually, or continuous chlorine monitors can be placed throughout the network. Use of continuous monitors is preferred since it will provide data at more frequent intervals than is feasible with manual sampling. Such a study is sometimes conducted in conjunction with a tracer test.

Case Study: Use Of A Tracer Study As Part Of A Master Plan

As part of a study to update the Master Plan and upgrade the hydraulic model for the area of Green River-Rock Springs, Wyoming, three tracer studies were conducted (Seppie et al., 2006). The tracer studies were designed to provide information for calibrating and validating the hydraulic model and to provide a better understanding of the movement of water through the distribution system. A plan for the tracer study was formulated and approved by EPA Region 8. Food grade calcium chloride was selected as the chemical tracer to be injected into the system. The quantity of chemical to be injected was calculated so that the EPA secondary MCL for chloride of 250 mg/L would not be exceeded. In preparation for the tracer study, the following activities were carried out: (1) develop a preliminary plan of action for the tracer study; (2) simulate the behavior of the tracer studies using the existing model of the distribution systems; (3) test the existing water quality in terms of normal background concentrations for calcium and chloride; (4) purchase the calcium chloride chemical; (5) acquire field equipment; and (6) develop a detailed operational plan.

Prior to the first tracer test in the northern portion of Green River, eight continuous conductivity meters (Figure 6-6) were installed at hydrants throughout the area. The tracer was injected using a variable speed pump (Figure 6-7) at a point at the upstream end of this zone and two valve positions were adjusted to ensure that no water entered the zone without going through the injection point. A clamp-on flow meter was installed downstream of the injection pump and flows were continually monitored and used to adjust the tracer injection feed so that the resulting chemical concentration stayed within an acceptable range. The tracer chemical was injected in the form of two 2-hour pulses with a period of approximately three hours between each pulse. The conductivity was automatically monitored for approximately 24 hours until it had fully spread throughout the northern Green River area. During each of the tracer tests, each meter was checked at least daily to ascertain that they were working properly and to download data. Additionally, at each meter site, manual chlorine measurements were taken at approximately three-hour intervals to provide data for use in the water quality portion of the Master Plan study.

Figure 6-8 is an example of the resulting conductivity records at two of the hydrant meters. The measured tracer concentrations were compared to the modeled results. The initial comparisons between measured and modeled concentrations varied among the meters. This data was later used to adjust model settings including pipe roughness factors and valve positions.

Source: B. Seppie, L. Kjellgren, W.M. Grayman. "Detailed Distribution System Modeling for a Regional Water Master Plan." Proc., Water Distribution System Analysis Symposium. University of Cincinnati, 2006.

Figure 6-6 Hydrant conductivity meter

Source: B. Seppie, L. Kjellgren, W.M. Grayman. "Detailed Distribution System Modeling for a Regional Water Master Plan." Proc., Water Distribution System Analysis Symposium. University of Cincinnati, 2006.

Figure 6-7 Tracer injection pump

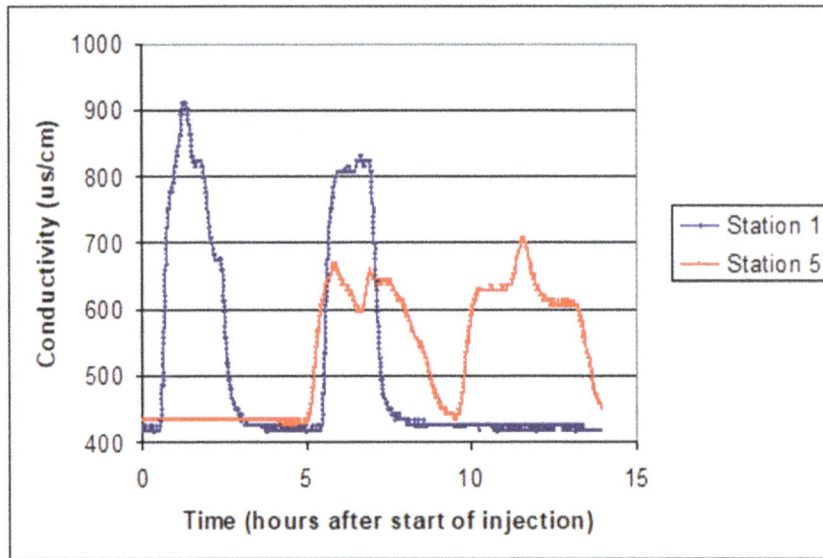

Source: B. Seppie, L. Kjellgren, W.M. Grayman. "Detailed Distribution System Modeling for a Regional Water Master Plan." Proc., Water Distribution System Analysis Symposium. University of Cincinnati, 2006.

Figure 6-8 Measured conductivity at two stations

REFERENCES

Monitoring and Modeling of the Sweetwater Authority Distribution System to Assess Water Quality. M. Hatcher, W.M. Grayman, C.D. Smith, M.A. Mann, *Proceedings, AWWA ACE.* 2004.

Detailed Distribution System Modeling for a Regional Water Master Plan. B. Seppie, L. Kjellgren, W.M. Grayman. Proc., Water Distribution System Analysis Symposium. University of Cincinnati. 2006.

Panguluri, S., W. M. Grayman, R. M. Clark, L. M. Garner, AND R. Haught. Water Distribution System Analysis: Field Studies, Modeling and Management. U.S. Environmental Protection Agency, Washington, DC, EPA/600/R-06/028, 2005.

Shang, F., J. G. Uber, and L. Rossman. EPANET Multi-Species Extension Software and User's Manual. U.S. Environmental Protection Agency, Washington, DC, EPA/600/C-10/002, 2011.

Reconstructing Historical Contamination Events. W.M. Grayman, R.M. Clark, B.L. Harding, M. Maslia, J. Aramini. Chapter in Water Security and Safety Handbook, Edited by Larry Mays, McGraw-Hill. 2004.

Janke, R., Tryby, M., and Clark, R.M. Protecting Water Supply Critical Infrastructure: An Overview. 2014. In Securing Water and Wastewater System: Global Experiences edited by Robert M. Clark and Simon Hakim. Springer-Science.

An Improved Water Distribution System Chlorine Decay Model Using EPANET MSX. Grayman, W., Kshirsagar, S., Rivera- Sustache, M., and Ginsberg, M. 2011. An Improved Water Distribution System Chlorine Decay Model Using EPANET MSX. In *Modeling of Urban Water Systems*, Monograph 20. Ed., W. James. Guelph, Ont.: CHI.

Clark, R.M. (2015). The USEPA's distribution system water quality modelling program: a historical perspective. Water and Environment Journal 29 (2015) 320–330.

Grayman, W.M., Clark, R.M., Males, R.M. (1988). Modeling Distribution-System Water Quality: Dynamic Approach. J.WRPMD, ASCE, Vol. 114, No. 3.

W.M. Grayman, L.A. Rossman, C. Arnold, R.A. Deininger, C. Smith, J.F. Smith, R. Schnipke (2000). Water Quality Modeling of Distribution System Storage Facilities, AWWA Research Foundation.

USEPA (1991). Water Quality Modeling in Distribution Systems. Proceedings. February 4–5, 1991.

J.J. Vasconcelos, L.A. Rossman, W.M. Grayman, P.F. Boulos, R.M. Clark. (1997). Kinetics of Chlorine Decay, , J.AWWA, Vol. 89, No. 7, July, 1997

Burkhardt, J.B, Szabo, J., Klosterman, J.S., Hall, J., Murray, R. (2017). Modeling fate and transport of arsenic in a chlorinated distribution system. Environmental Modelling & Software, Volume 93, July 2017, Pages 322–331.

Helbling, D.E, and VanBriesen, J. (2009). Modeling Residual Chlorine Response to a Microbial Contamination Event in Drinking Water Distribution Systems. ASCE Journal of Environmental Engineering 135(10):918–927

AWWA, 2018. Computer Modeling of Water Distribution Systems. Manual 32.

AWWA, 2017. Water Quality in Distribution Systems. Manual M68.

CHAPTER 7

Calibration for Energy Management

Tom Walski

Hydraulic models are being increasingly used for studies of pumping energy to identify opportunities for energy and cost savings and more efficient system operation. Before using a model for an energy study, it is important to determine how well that model can predict energy usage and ultimately, energy cost.

Because efficiency and cost data are not needed for a typical water distribution system model, a model may be well calibrated for a standard extended period simulation (EPS) run (or even a water quality run) but do a poor job in predicting energy use. Some additional work is needed to transform a calibrated hydraulic model into a calibrated energy model.

There are many benefits from calibrating a model for energy usage. During energy calibration, errors in the underlying hydraulic model may be identified. Insights derived from calibration often make it easy to identify areas where energy or cost savings can be achieved. For example, pumps operating at poor operating points or questionable decision-making by operators can be found. In reviewing energy cost, it may also be possible to discover billing errors by the power utility or find situations where a more beneficial energy rate tariff can be uncovered. Reducing energy use is described in many publications including (AWWA, 2016; Senon, et al., 2015; WEF, 2010; Walski, 2011) and modeling is a valuable tool for these calculations. It is also useful for determining future energy usage for projected demands or changes to system operations, such as pressure zone shifts, the addition of water storage, or re-configuring the use of a pump station that encounters high-demand charges but low power usage.

There are actually several facets to energy calibration. First, it is necessary to decide if the calibration is to be based on energy use or energy cost. Predicting energy use is somewhat of an easier task since it doesn't involve energy tariffs

---

Calibrating a Model for Energy

A large regional system spent roughly 20 percent of its operation budget on energy, and they realized that the model could help them identify opportunity for savings.

A smart first step was the realization that the quality of energy results from the model depended on the accuracy of the underlying EPS model. They worked hard to get their EPS model to accurately match actual flows and pressures at the pump stations.

Once the EPS calibration was very good, they were able to bring in power data for each pump station and match power usage in the model with pump station measured use. The major issue was that power was measured at the electric utility meter level, not at individual pumps, so adjustments had to be made to calibration data to account for other power users on the meter. With that, they were able to produce graphs such as the following.

---

which can be complex. However, calculating energy use correctly does not guarantee that energy costs can be predicted accurately.

The next facet is the temporal scale of the calibration. Will the model be used to calculate instantaneous (or short-term) use, daily use, use over a billing period (usually month), or long-term energy use (which can involve seasonal trends)? How will anomalous events such as fires, large pipe breaks, or special events be accounted for in the calibration and subsequent modeling?

Finally, there are spatial considerations. Will the calibration be performed for individual pumps, pump stations, aggregate power meters/energy bills or system wide?

Many of the concepts in this chapter apply to energy generation at turbines as well as energy use at pumps but the calibration for energy use will be presented because pumps are much more common than turbines in water distribution systems.

This chapter begins with a description of additional data collection for energy calibration; then the basics for energy calibration and the background on how models calculate energy use are described; determination of how to adjust a model to achieve energy calibration is presented; and finally, special issues in converting from energy cost to energy use are described.

DATA REQUIRED FOR ENERGY CALIBRATION

Because energy use and energy cost are the basis for judging energy calibration, energy use data for pumps (and energy cost data if cost comparisons are used) must be collected. Generally large pumps at the main treatment plant or large pumping have the most extensive energy data while pumps at small remote stations or wells have the most limited data.

In some systems, the SCADA (Supervisory Control and Data Acquisition) or plant control system collects and stores a great deal of useful data. In some situations, the power utility also collects considerable data on energy consumption and is willing to share it with the water utility. However, power utility data is usually aggregated spatially to the level of the power meter or account and not down to the individual pumps or other uses.

Ideally, each pump will have its own power meter which will have a history of power usage time stamped so that it can be matched with pressure and flow data. In other cases, however, only instantaneous readings are available at the site with a meter. In many, energy use is not recorded and other methods must be used. Installing a temporary power meter or ammeter with data logging may be needed. In some cases, it may be easier to install a temporary ammeter and calculate the energy use from the amp draw. Finally, the station's cumulative energy use may be read over a time period from the building's power billing meter during a time when the number of pumps running is constant. Otherwise, there is too much uncertainty in which pumping condition corresponds to the energy use.

In some installations, apparent power (in kVAR) rather than real power (in KW) is reported. In DC motors, these will be the same, but in AC motors, they will differ by the power factor which depends on the phase angle according to kW = kVAR x Power factor.

In reviewing energy use data from the SCADA system, it is important to understand how time was considered because SCADA systems only report data at a "polling interval" which may be on the order of seconds to hours. In addition, SCADA historians may perform some manipulation of the data such as reporting values as an hourly average. The modeler needs to understand if the reported values represent instantaneous value at the time stamp (in kW), the average during the previous polling interval (in kWhr), or some other value. If the pump

Tracking Pump Station Energy Use

A large water utility had two pump stations serving a relatively remote section of their system. They recently made modifications to the system and wanted to check if the pumps were operating efficiently.

They used a calibrated model to compare energy actually used with the model prediction. The calibration consisted primarily of checking the EPS with regard to pump operating points and tank level fluctuations. The model also agreed with the energy billing data.

Moreover, it showed that the pumps were operating at inefficient operating points such that instead of operating at roughly 80 percent efficiency, they were operating near 50 percent efficiency. Recommendations to reduce energy consumption were developed.

changed status during an averaging period (e.g. on/off, change in pump speed), it is important to understand the nature of the change.

If power meters are used instead on individual pump data, it is necessary to understand all of the power uses associated with the power meter. These may include energy used for lighting, heating, SCADA equipment, chemical feed equipment, and other pumping (e.g. sump pump, seal water pump). In most cases, these will be small values and can be estimated and subtracted from the total energy use to arrive at pumping energy. However, at the main pump station located at a treatment plant, raw water pumping, treated water pumping, and backwash pumping may all be included in a single power reading. This problem is exacerbated in installations with large energy using treatment processes such as reverse osmosis and ozone generation. In such cases, separate metering for distribution pumping is required because estimating these large non-pumping uses will be inaccurate.

ENERGY MODELING BASICS

Calculating energy use and cost represents an extension to standard hydraulic model calculations which have as its goal calculation of hydraulic properties such as flow, pressure, and velocity. To extend the calculations to energy use, it is necessary to have data on pump and motor efficiency (or power) as a function of flow and pump head. To further extend energy use results to energy costs, it is necessary to have data on pumping unit prices which are usually not a simple constant unit price in most instances.

There are two overall types of energy models:

1. Calculating energy use and greenhouse gas emission
2. Calculating energy cost

The first is the easier and a prerequisite for the second. Starting with a calibrated EPS model, the only additional data needed to determine energy consumption is the efficiency of the pump and motor. It is important to distinguish between pump efficiency and wire-to-water (overall). The pump efficiency is usually given in manufacturer curves and resents the conversion of motor power to water power. The second represented the conversion of electric power to water power and is usually what is measured in the field.

Models to calculate energy cost are more complex as they must also know the energy cost tariff which can be quite complete and include such factors as time-of-day pricing, peak demand charges, and seasonal pricing among other factors. Data to calculate actual energy costs can often be provided by the power utility.

While energy use can be calculated based on individual pumps or pump stations, cost estimating must usually be done at the level of pump stations because costs are not calculated by pump.

The cost of energy over some period of time can be given by:

$$C = k\gamma \sum_{i=1}^{N} \frac{Q_i h_i p_i \Delta t_i}{e_{pi} e_{mi} e_{di}}$$

Where C = cost of energy between times 1 and N; k = unit conversion factor; γ = specific weight of water; Q_i = flow at time step i; h_i = pump head at time step, i; p_i = price of energy at time step i; Δt_i = length of timestep i; e_{pi} = pump efficiency at time step i; e_{mi} = motor efficiency at time step i; e_{di} = variable speed drive efficiency at time step i.

In some cases, the unit price of energy may be based on block rate pricing where the unit price is based on total and/or peak energy use and is not known until the end of the billing period.

In addition to energy cost, many energy tariffs contain an additional cost for a demand charge based on peak energy use during some period of time. Formulation of demand charge can vary widely between tariffs and can depend on the time period for which it is calculated (e.g. summer only, coincident peak period), the length of the peak usage (15 minutes, hour) used in the calculation, and the persistence of the demand charge once it is set (current billing period, one year).

With the advent of AMI, it has become easier to obtain power consumption and cost information, sometimes in near real time.

MODEL DATA REQUIREMENTS

The calibrated hydraulic model calculates pump head and flow values. The key additional input for energy calculation is efficiency of the pump, motor, and any variable frequency drive (if applicable). Pump efficiency is a function of flow and is available from manufacturer test data. Motor efficiency is available from the motor manufacturer and is relatively constant unless the power use drops far below (<50%) of the rated power of the motor. Variable speed pump drives are usually based on variable frequency drive (VFD) technology and data on VFD efficiency, which varies with motor speed, is available from most manufacturers although it may be more difficult to obtain than the other efficiencies.

Manufactures' data can be used for pump, motor, and VFD efficiency. However, it is desirable to check efficiency in the pump station. Measuring the individual pump and motor efficiencies in the field is often difficult and most of the time, only the overall efficiency (also called wire-to-water efficiency) is measured. This is the ratio of power added to the water divided by power measured at the power meter (i.e., $k\gamma Qh$/(power in)).

Measuring flow and pressure is described earlier in this publication. Input power measurements represent an additional measurement to determine efficiency. Procedures are described in electrical references such as IEEE (2014).

Efficiency curves, especially pump efficiency, can vary over time as pump performance deteriorates. Efficiency values should be checked and field tests of efficiency may reveal the source of error in model energy calculations or pump mechanical problems. When measuring pump efficiency, it is best to attempt to measure it over as wide of a range of flows as possible rather than at a single flow rate (Figure 7-1).

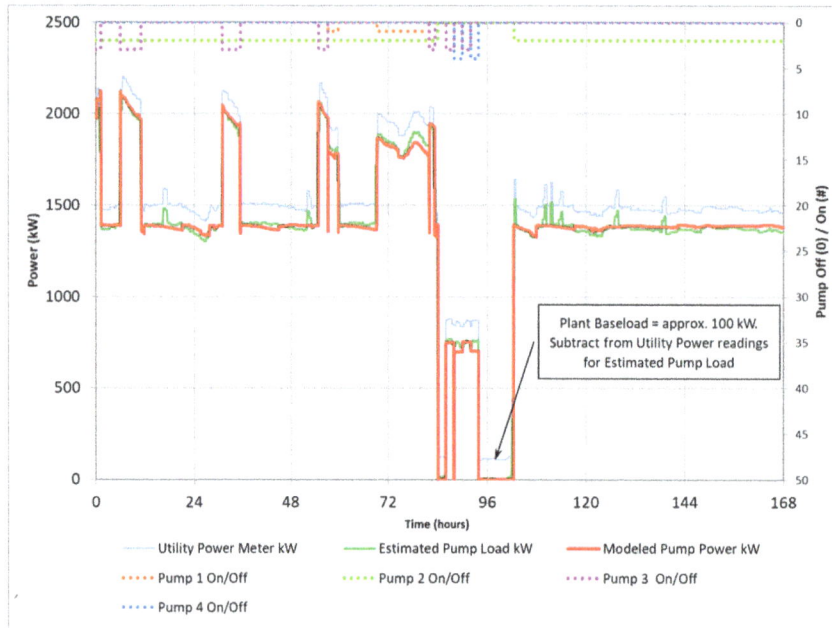

Source: Courtesy of Cobb County-Marietta Water Authority

Figure 7-1 Illustration of power usage in the model matched with pump station measured use

ADJUSTMENTS TO ACHIEVE SUCCESSFUL CALIBRATION

As with hydraulic calibration, when the model calculations differ from field data, it is necessary to determine the cause of the discrepancy and correct it. This is done by isolating individual sources of error and testing to see if adjustments can bring the calculations into agreement.

Before beginning to perform energy calculations, it is important that the EPS hydraulic model is well calibrated. If it is inaccurate, then energy calculation will reflect an inaccuracy. The most common source of discrepancies in EPS modeling arises when operators do not operate the system according to the control rules contained in the model. While the control rules are needed for predictive runs of the model, it may be necessary in calibration runs to override those rules to reflect historical operation for the time for which energy use data is available. Other potential sources of error when calibrating for energy usage, in order of likelihood, include controls, demand (and patterns), efficiencies, and field data accuracy.

The model runs used in calibration must be based on the demands for the time period (e.g. day) on which the energy measurements were made. Models are often calibrated for an average day but modifications to the base demand or the demand pattern may be required for the calibration time period. It is important to correct for any anomalous demands during the testing period such a as fires, large pipe breaks or special events (e.g. sporting events, concerts, holidays).

Once the system hydraulics are in agreement with the test day's measurements, the next most likely source of error is in the efficiency curves. Wire-to-water efficiency test, described in the previous section may be needed to verify the accuracy of the original efficiency curves. For constant speed pumps, the efficiency varies with flow but there is only a single curve. For variable speed pumps, it is essential to record the actual pump speed corresponding to energy usage as the efficiency curves themselves vary with speed. While collecting data for variable speed pumps, it may be necessary to override the automatic feedback loop that controls speed and fix the speed while measurements are made. The range of speeds should correspond to the range found during normal operation.

As with hydraulic calibration, the accuracy of field data must be verified. Flow and pressure data should already be accurate if the hydraulic model has been well calibrated. Power meters should be tested if there is any chance that they may be in error. If relying on the metering provided by the power utility, it may be necessary to have a discussion with them about the resolution and metering accuracy for their data.

ENERGY COST CALCULATIONS

The most rigorous test of energy model calibration is the ability of the model to reproduce energy bills from the power utility. If the energy cost was simply the total energy use times the unit price of energy, then this would be a simple step. However, energy tariffs are notoriously complex and hence there are numerous sources of error in converting energy use into energy cost, including time-of-day pricing, block rate pricing, seasonal pricing, demand charges, and purchasing energy on the open market. (Walski and Hartell, 2012)

The modeler must obtain the current energy tariff from the power company and understand it, which can be difficult. The modeler should also collect energy bills from several billing periods to understand how the tariff is applied to the water utility. Correct matching the energy bill is the target of calibration for energy cost.

In calibrating pumping energy costs, it is first necessary to subtract fixed charges and non-pumping energy charges from the energy bill, resulting in only pumping energy cost.

Once the non-pumping energy costs have been removed from the bill, the energy cost and demand charge need to be separated and analyzed individually. A simple approach is to calculate energy cost for an average day and multiply it by the number of days in a billing period. This may be adequate in some cases but does not account for the fact that there may be dry and wet days, weekend and weekday, and days with special events in the billing period. Since no two days are exactly alike, a full simulation of the entire billing period may be more accurate but require a good deal of setup to make such a run.

A compromise may be to run several typical days (dry weekday, wet weekend) and take a weighted sum of the energy used on each day and then apply that amount to energy costing. This works well for block rate tariffs. For time-of-day pricing, it is necessary to calculate the cost for each day and take a weighed sum of the costs and not of the usage.

Demand charges are not determined based on cumulative energy use but on the peak energy usage during some specified time period (or over the entire billing period or year). First, review the tariff to understand if the period over which peak energy usage is measured is a 15-minute or hour period or some other duration. Also determine if the peak demand is calculated system-wide or individually at each power meter.

The modeler must find the time period for which the peak demand was established and simulate the hydraulic conditions in the system at that time. The modeler must estimate the demand at that time, the tank levels, and which pumps were operating. A short EPS run should be made for the peak energy use period and that energy use and demand charge in the model should match the billed peak demand. If not, as with any other model, the modeler must determine the reason for the discrepancy and correct it. Reasons can include incorrect demand (e.g., a fire, or intermission at a concert or sporting event), incorrect pump status (e.g., some backup pump have been tested), or incorrect efficiency curves (e.g., pump have lost efficiency over time). It may be necessary to meet with the power utility to understand exactly how they perform the demand charge calculation.

REFERENCES

AWWA, 2016, *Energy Management for Water Utilities*, American Water Works Association, Denver, Colo.

Institute of Electrical and Electronics Engineers, 2014, *Standard Test Procedure for Polyphase Induction Motors and Generators*, IEEE 112-2004,IEEE Press, Los Alamitos, Calif.

Black, J., 2018, "Hydraulic Modeling and Energy Management," *AWWA Water Infrastructure Conference*, Denver, Colo.

Senon, C., et al., 2015, *Drinking Water Pump Station Design and Operation for Energy Efficiency*, Water Research Foundation, Denver, Colo.

Walski, T., 2011, "Practical Tips for Reducing Energy Use," *Computers and Controls in the Water Industry Conference*, Exeter, UK, Sept 2011.

Walski, T., and Hartell, W., 2012, "Understanding Energy Pricing for Water Pumping," *Water Distribution System Analysis Conference*, Adelaide, Australia.

WEF, 2010, *Energy Conservation in Water and Wastewater Facilities*, Water environment Federation Manual of Practice No. 32, Alexandria, Va.

INDEX

Note: *f* indicates figure; *n* indicates (foot)note; *t* indicates table.

www.ingramcontent.com/pod-product-compliance
Lightning Source LLC
Chambersburg PA
CBHW081545220326
41598CB00036B/6567